T0267314

THE
CRANBERRY
HARD WORK AND
HOLIDAY SAUCE

Hand harvest, Harwich, Massachusetts.
Harwich Historical Society

Wet harvest, Maple Springs Bog, Wareham, Massachusetts.
Lindy Gifford photograph

The Cranberry

Hard Work and Holiday Sauce

Stephen Cole
and Lindy Gifford

Camden, Maine

To the memory of

Alton H. Cole

Nathaniel H. Gifford

Charles W. Harris

Malcolm E. Ryder

Down East Books

An imprint of Globe Pequot

Distributed by NATIONAL BOOK NETWORK

First paperback edition: 2009
First Down East Books edition: 2023

Library of Congress Cataloging-in-Publication Data

Cole, Stephen A., 1955-
The cranberry : hard work and holiday sauce / Stephen Cole, Lindy Gifford. – 1st paperback ed.
p. cm.
Includes bibliographical references and index.
ISBN 978-1-68475-127-3 (pbk. : alk. paper)
1. Cranberries. 2. Cranberries--United States--History. I. Gifford, Linda S. II. Title.
SB383.C65 2009
634'.760973–dc22
2009022890

Designed by Lindy Gifford, Damariscotta, Maine.
Copyedited by Genie Dailey, Finepoints Editorial Services, Jefferson, Maine.

∞™ The paper used in this publication meets the minimum requirements of American National Standard for Information Sciences-Permanence of Paper for Printed Library Materials, ANSI/NISO Z39.48-1992.

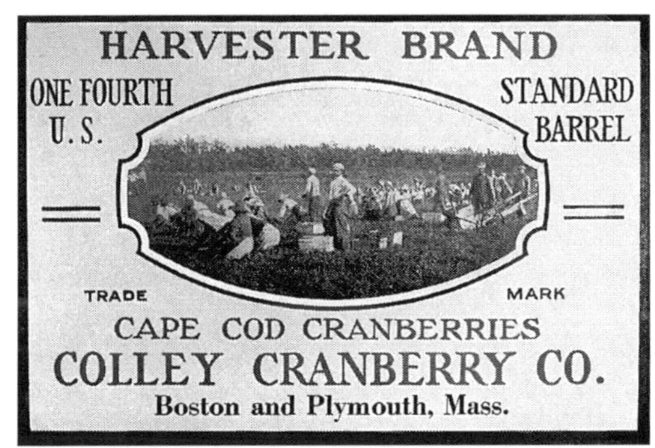

Cranberry shipping box labels.
Middleborough Public Library & private collections

Contents

Philip Brackett, Chopchaque Bog, Mashpee, Massachusetts.
Lindy Gifford photograph

Amy Carleton, Augusta Carleton Jillson, and Annie Carleton Lloyd, East Sandwich, Massachusetts. Lindy Gifford photograph

Acknowledgments

Books are collaborative efforts, so there are always people to thank for their help. The writer's great uncle, Charles W. Harris, made contacts and suggestions from a lifetime spent in agriculture. Years before, he had given his grand nephew an introduction to cranberry growing via an irrigation system that needed installation. Chris Makepeace allowed us to roam the sand roads, bogs, and uplands of the A. D. Makepeace Company, and Bob Conway, foreman of the Makepeace country known as Swan Holt, provided insights, asides, and a job on the dry harvest. Philip Brackett—whose lovely, well tended Chopchaque Bog made us want to become cranberry growers—was a regular host and friend.

Part of the cranberry's story lies in libraries and archives throughout the Commonwealth of Massachusetts. The staffs of the Middleborough Public Library (keepers of the Cranberry Room), Cape Cod Community College, and the Concord Public Library were especially helpful in our search. (It's possible to write Middleborough or Middleboro, and when given the option, we've chosen the latter, as many locals have.) At the UMass Cranberry Station in East Wareham, researchers allowed access to the documents and photographs collected there. We are also indebted to Skip Colcord of Ocean Spray Cranberries, Inc., for opening the company's photo archives to us. History resides in newspapers as well, and we are grateful to the *Wareham Courier* staff for allowing us to rummage through their back pages.

Several faculty at Brown University arranged for technical assistance to this project. Bruce Rosenberg and David Marc of the Program in American Civilization provided computing time on the university's mainframe. At the Department of Anthropology, Douglas Anderson put a darkroom at our disposal. Finally, we are happy to acknowledge the aid and support of our advisors: anthropologist Nora Groce, geographer Arthur J. Krim, and the late cranberry expert Irving E. Demoranville. Time and again they observed, read, critiqued, and encouraged.

We are also indebted to the following individuals, living and departed, who shared with us both time and recollections: Augusta C. Jillson, Annie E. Lloyd, and Amy Carleton, East Sandwich, MA; Mary Peterson Simpson, Harwich, MA; Richard Kiernan, Wareham, MA; Doris S. Gomes, Marion, MA; Malcolm and Katherine Ryder, Cotuit, MA; Robert C. Hammond, Wareham, MA; Antone J. Jesus, Onset, MA; William E. Crowell, Dennis, MA; Lawrence Cole, Carver, MA; Edward B. Garside, Plymouth, MA; Wilho and Lillian Harju, Carver, MA; W. Marland Rounsville, Nantucket, MA; Ellen Stillman, Hanson, MA; Maurice B. Makepeace, Marion, MA; Philip Brackett, Cotuit, MA; Eunice A. Bailey, South Carver, MA; Jennie Bailey Shaw, South Carver, MA; Gilbert T. Beaton, Buzzards Bay, MA; Vincent and Beatrice Pina, Marion, MA; Leonard F. Vanderhoop, Sr., Gay Head, MA; Raymond L. Thatcher, Harwich, MA; Ernest D. Howes, Wareham, MA; Orrin G. Colley, Duxbury, MA; and Sylvester Andrews, Falmouth, MA.

Finally, thanks—yet again—to Jennifer Bunting and Tilbury House, Publishers, for scooping up a manuscript others had left behind.

This volume was prepared, in part, with support from the National Endowment for the Humanities, an independent federal agency.

Stephen Cole & Lindy Gifford

Locke Bog, Wareham, Massachusetts.
Lindy Gifford photograph

Introduction

Entering Cranberry Country

In the last few miles above the canal, the summer visitor to Cape Cod passes through towns called Middleboro, Carver, and Wareham. Always eager to reach the Cape, the traveler will never know these places as anything more than roadside signs. But there is life here, and life of a very particular sort, for beyond the highway are 11,000 acres of bog, ditched and vine-covered. Each fall, after the last tourist has gone home, men appear on the bogs with machines suggesting great riding eggbeaters to harvest a third of the nation's cranberries, the turkey's tablemate.

There was a time when the traditional crop of Cape Cod and Plymouth County made itself plain to summer people. In the 1950s, the tourist coming from Boston or its suburbs traveled Route 28, the Cranberry Highway. In South Middleboro, the family could stop at cranberry grower Ben Howes's Chicken House, where they were served cranberry sauce no matter what was ordered. Further along the road, in Onset, the motorist pulled up to The Bottle, an outsized juice container from which cranberry cocktail and delicacies such as cranberry sherbet were sold. The smaller, older highway ran closer to the bogs. On a summer day, the grower might be seen cutting grass around the edge or standing knee-deep in water, clearing the ditches of plants and muck. Rarely would a figure appear on the bog itself. The vines were in bloom, so the grower confined his activities to the periphery and let the honeybees work the bog.

Today, once the traveler crosses the Cape Cod Canal, there are fewer bogs to be seen. Though it was on Cape Cod early in the 1800s that vines were first taken from the wild and encouraged, before the century was out the bulk of cranberry growing was done in Plymouth County, where the cedar and maple swamps for building bog were more numerous. Since then, Cape Cod cranberry growing has further declined. Since the close of World War II, bogs have rapidly been displaced by motels, campgrounds, vacation homes, and mini-malls. "Cape Cod Cranberries," stenciled on a cellophane bag at the supermarket has become a misnomer. But the symbolic link between Cape Cod and the cranberry remains strong. For years Gerry Studds, the region's congressman, kept cranberry juice in his Washington water cooler. And scattered throughout the towns were enterprises called Cranberry Cottages, Cranberry Moose Restaurant, Cranberry Cleaners, Cranberry Home Protection Service, Cranberry Inn, Cranberry Muffin, Cranberry Travel, Cranberry Valley Golf Course, and Cranberry Corner.

Driving back from Cape Cod into Plymouth County, taking any exit will make its agricultural identity plain. Pickup trucks pass with license plates declaring this "Cranberry Country." The berry is so important to rural Carver that it appears on the town seal. A dirt road will inevitably lead to a bog, vast and rectangular or small and fitting the contours of the land. In this place, there were for years more acres of cranberry bog than people. In the words of a Carver native, "Cranberries was all there was."

In little more than a century, this diminutive fruit has forced those who would profit from it to become inventive and resourceful, to make for it a home, to protect it from the wind and cold, and to fashion tools to tend and pick it. For some, the cranberry has provided a comfortable living for five generations, for others, only enough money to make it through each winter. When something so dominates the lives of people, it is worth knowing about.

A patch of wild cranberries. Sandy Neck, Sandwich and Barnstable, Massachusetts.
Lindy Gifford photograph

Chapter 1

Wild Harvests

They now began to gather in the small harvest they had, and to fit up their house and dwelling against winter, being all well recovered in health and strength and had all things in good plenty. For as some were thus employed in affairs abroad, others were exercised in fishing, about cod and bass and other fish, of which they took good store, of which every family had their portion. All the summer there was no want; and now began to come in store of fowl, as winter approached, of which this place did abound when they came first (but afterward decreased by degrees). And besides waterfowl there was great store of wild turkeys, of which they took many, besides venison, etc.

So reads William Bradford's diary on preparations among the Pilgrims at Plimoth for the very first Thanksgiving. For this three-day feast the cod and bass were perhaps poached in a broth seasoned with thyme and rosemary. Venison, contributed by the native Wampanoags, was parboiled and then roasted, as were the ducks and geese. Turkey came to the table roasted, or as the middle in a great pie. A coarse cornbread and the Pilgrims' ubiquitous stewed pumpkin complemented the meats, all washed down with homemade beer. There may have been fruit tarts to enjoy, made from dried wild strawberries or grapes. But where in the menu came cranberries, the turkey's companion, our traditional holiday fruit so symbolic of Thanksgiving and things American? The truth is that if the cranberry made an appearance at all, it did so as a minor ingredient in stuffing for the roast birds. The cranberry had to wait for the importation of sugar to New England before taking its place beside the turkey as part of America's traditional cuisine.

The Indian peoples, who were reluctant hosts of the English in coastal New England, had no qualms about using the cranberry as nature had made it; and there is even doubt whether sweetness was a taste that appealed to them. Natives used the berry in two ways, as medicine and as food. When Roger Williams—banished to Rhode Island—wrote his observations on the habits of the native Narragansett, *A Key to the Language of America,* in 1643, he included the word "sasemineash." Williams knew no English equivalent for the word, but described sasemineash as a "sharp, cooling Fruit growing in fresh Waters all the Winter." Sasemineash was, he commented, "excellent in conserves against Fevers." Conserves and preserves were considered medicines in the seventeenth century. Cranberries, as well, were used in poultices to heal sores and draw infection from the flesh. Unknown to the users, it was the presence of iodine, calcium, iron, and vitamins A, C, and B that made the berry valuable in healing the sick. So revered were the fruit's qualities that an Indian is recorded as having made a journey of several days in search of cranberries to ease his ailing wife.

As part of the Native diet, wild cranberries, picked in the autumn or winter, became a sauce to be eaten with meat. This is the way explorer Marc Lescarbot enjoyed them, while observing the Indians of Maine and the Maritime Provinces in the early seventeenth century. Ripe cranberries may also have studded pemmican, the pounded and dried strips of meat and fat sometimes peppered with wild berries. Drying the many wild fruits the land offered besides the cranberry—grapes, beach plums, gooseberries, wild cherries and strawberries, raspberries, crab apples, currants, and blueberries—was a common means employed by Indians to store them for year-round use. Dried strawberries might adorn a bread made with cornmeal in midwinter; currants and cranberries, both dried, would enliven the cornmeal mush called samp. It seems almost certain that the Native dwellers of Massachusetts and Rhode Island—the Wampanoag, Massachusett, and Narragansett tribes—introduced the cranberry to the English, as they did the vegetables they cultivated: corn, pumpkin, and squash. What the newcomers saw in the wetlands was best described in the mid-seventeenth century by the British traveler and amateur botanist John Josselyn, who scoured New England in search of its rarities.

Cranberry or bearberry (because bears use much to feed upon them), is a small, trayling plant that grows in saltmarshes that are overgrown with moss. The tender branches, which are reddish, run out in great length, lying flat on the ground, where, at distances, they take root, overspreading sometimes half a score acres, sometimes in small patches of about a rod or the like. The leaves are like box, but greener—thick and glistening. The blossoms are very like the flowers of our English nightshade; after which succeed the berries, hanging by long, small, footstalks, no bigger than a hair. At first, they are of a pale yellow colour; afterwards red and as big as a cherry: some perfectly round, others oval; all of them hollow; of a sour, astringent taste. They are ripe in August and September.

Perhaps most interesting in Josselyn's description is that he wrote of the cranberry as if it did not exist at home. In fact, a kind of cranberry does grow in the British Isles, doing best in the north of England and in Scotland, though it was not widely known or used prior to the discovery of an American cranberry. The difference between these varieties, Great Britain's *Vaccinium oxycoccus* and the North American native, *Vaccinium macrocarpon,* is mostly one of size. The small cranberry found in northern Europe, Russia, and North America, as well as in the British Isles, is about the size of a pea. The large cranberry, growing only on the North American continent, is a berry a full half-inch in diameter. Both plants are heaths, part of a family that includes other evergreen shrubs such as mountain laurel, wintergreen, trailing arbutus, heather, and the North American natives, high- and low-bush blueberry. In places where moderate weather prevails and soils are acid—on moors, dunes, mountain slopes, and in swamps—the heaths are common vegetation. What sets the large and small cranberry plants apart from other heaths is their need for a great deal of water. This characteristic is attested to in the host of names by which cranberries were known in England: marsh-whorts,

The American cranberry from William Aiton's *Hortus Kewensis*.

fen-whorts, fen-berries, marsh-berries, and moss-berries. They are names, however, that rarely appear in British cookbooks of the late sixteenth and early seventeenth centuries.

As befits a native fruit, the name cranberry (and its variants craneberry, cramberry, and craberry) evolved in North America, the word traveling to Great Britain along with imported berries by the 1680s. Its appearance came between Williams's 1643 writings, when sasemineash had no known English translation, and 1651, when Plymouth Colony records provide the first written mention of "cramberries." The name derived, the folk explanation goes, from the appearance of the shrub in bloom. Its beak-like stamen, swept-back petals, and drooping stalk resembled the neck and head of the crane, which still inhabited New England's freshwater wetlands in the seventeenth century. The cranberry is not unique among plants in taking a name from the bird-like shape of its blossom. The flower of the geranium also resembles a beak and is sometimes called crane's bill, stork's bill, or heron's bill.

It was the beginning of trade with another English colony, the West Indies, that brought sugar in quantity to New England and put the cranberry on the map, making it a traditional fruit for cooking. In 1647 Governor Winthrop of Massachusetts Bay wrote of New England's pleasure at the new trade with the Indies. Sugar first came from cane plantations on Barbados, then from Jamaica and the Leewards. Before this time, there was simply no reliable source of sweetening in southern New England. The Natives did not slash maples and collect the sap as was common among Indians in northern New England, nor did wild bees produce honey sufficient for collecting. When the English began importing the honeybee at mid-century, the Indians referred to them as "English flies."

British cookbooks of the early seventeenth century, such as John Murrell's *A New Booke of Cookerie* and Gervase Markham's *Country Contentments or The English Huswife,* are filled with recipes in which fruits of many kinds garnished roasts, were tossed into puddings, or were baked in a crust. Barberries ornamented boiled pigeon and rabbit. Prunes, raisins, and currants were boiled with meat to make a broth. Puddings were not complete without a sprinkling of currants. And every fruit possible was mixed with wine, sugar, cinnamon, and ginger to make a filling for tarts. Once the cranberry was made palatable, it is not surprising that it quickly became popular among the settlers of New England.

The fruit also became a prominent entry in British commentaries on New England. The agriculturist Samuel Hartlib, writing in 1655, praised from afar the cranberry's value to the body and the palate. "They are accounted very good against the scurvy, and very excellent in tarts," he wrote. "I know not a more excellent and healthful fruit." John Josselyn, who described the plant so carefully in his *New England's Rarieties Discovered* of 1672, added a note-like postscript on the berry's singular qualities:

> For the Scurvey. They are excellent against Scurvey. For the Heat in Feavors. They are also good to allay the feavor of Hot Diseases. The Indians and English use them much, boyling them with Sugar for Sauce to eat with their Meate, and it is a delicate Sauce, especially with Roast Mutton. Some make tarts with them as with Goose Berries.

To discover a fruit that was tasty, yet medicinal, amid the doubtful medical practices of the seventeenth century, was clearly a bonanza.

With the cranberry's growing popularity was begun a practice Native Americans had long carried on, autumn picking of the wild fruit amid the bogs and wetlands of the great New England wilderness. It would be a tradition enjoyed yearly in an unbroken chain to this day. All throughout the late seventeenth and eighteenth centuries, the cranberry continued its insinuation into the seasonal traditions and diet of New England and the mid-Atlantic. In 1796, when the country's first cookbook, Amelia Simmons's *American Cookery,* appeared, it could already be suggested that when roasting turkey, "serve up with boiled onions and cranberry-sauce, mangoes, pickles, or celery."

The degree of entrenchment achieved by the native cranberry within American cuisine is best seen in a satirical essay titled "Cranberry Sauce" in 1808 by the Anthology Society, later to become the hoary Boston Athenaeum. Its author was William Tudor, who, beyond this spoof, could count among his accomplishments the notion of selling New England ice in the West Indies, a proposal for the erection of Bunker Hill Monument, and coinage of the phrase "Athens of America" to describe Boston. In the essay, Tudor posed as a citizen of France who observes with no end of loathing gastronomic preparations in the new republic. Among the litter of puddings composed of "heterogeneous materials" with a specific gravity "greater than that of lead" and vegetables "boiled and soaked in hot water," Tudor focused his attention on cranberry sauce, ("vulgarly called cramberry sauce, from the voracious manner in which they eat it").

Cranberry sauce was, to Tudor's mind, our "most universal dish, which obtains equally at the tables of the rich and poor.... Preparing them for the table is very easily done," he explained:

> ...the berries are stewed slowly with nearly their weight of sugar for about an hour, and served on the table cold; the sugar made use of differs in quality according to the wealth of those by whom the sauce is used. It is eaten with almost every species of roasted meat, particularly the white meats, turkies, partridges, &c. Some even eat it with boiled fish, and I knew one person, otherwise a very worthy man, who eats it with lobsters for supper.

The peculiar qualities of the cranberry and its sauce were the chief impediment to Americans giving up "huge pieces of half boiled meat, clammy puddings, and ill-concocted hashes" in favor of the cuisine of France, where food and its preparation had been raised to the level of art.

> From the most accurate observations, I am convinced, that French cookery, to which they generally have a dislike, will never be effectually introduced among them, till the preparation treated of in this memoir, shall be no longer used; because, from its universal use, possessing a mixture of sweetness and acidity, it stimulates their appetite, and prevents them from perceiving the insipidity and staleness of their dishes, and makes them insensible to the advantages of our various rich sauces.

William Tudor was realistic about the prospects of weaning Americans from cranberry sauce.

> The difficulty attendant on the achievement of this reformation cannot be concealed, the custom is universal. Dining once with one of the cabinet ministers at the seat of government, there were four soup plates of this article, at the four corners of the table, which is strong proof that this practice is carried on by persons even in the most exalted stations, though he was from the portion of the United States where the habit is most inveterate.

Perhaps what made the cranberry so popular in America was the berry's identity—it too was American. The largest, most robust cranberry did not grow in Europe, it thrived only on our continent.

It might be suggested...that, by disuse of this fruit, a large quantity of meadows, now useless, might be reclaimed...that is now lost by suffering the growth of this pernicious berry, which, in its preparation requires such a quantity of sugar, as to greatly increase their humiliating dependency on the colonies of foreign nations.

—William Tudor, "Cranberry Sauce,"
William Tudor: Miscellanies

As sauce, it was an American delicacy, not a dish borrowed from the Old World.

But the force of Tudor's argument against the "preparation" called cranberry sauce rested not only on culinary aesthetics, but also on economic advantages, as he wryly pointed out in conclusion:

> It might be suggested further to their political economists, that, by disuse of this fruit, a large quantity of meadows, now useless, might be reclaimed and added to their national resources: that a very considerable addition of wholesome food would be thus procured for their horses and cattle, that is now lost by suffering the growth of this pernicious berry, which, in its preparation requires such a quantity of sugar, as to greatly increase their humiliating dependency on the colonies of foreign nations.

Whatever William Tudor's true feelings about the small fruit he had so maligned in print, it is obvious that this native berry had become firmly established in the national gastronomy at an early date. The cranberry was now a traditional dish and to one acute observer, symbolic of things distinctly American.

Sudbury River meadows.
Courtesy Concord Free Public Library

Barrett shows me some very handsome pear-shaped cranberries, not uncommon, which may be a permanent variety different from the common rounded ones.

—Henry D. Thoreau,
The Journal of Henry D. Thoreau

Despite the work of the satirist, cranberries were eaten in new ways, and not less, but more, as the nineteenth century wore on. Almost every American cookbook written introduced a novel cranberry recipe. The 1807 *A New System of Domestic Cookery* suggested stewing cranberries with sugar and putting them up in jars, "which way they eat well with bread and are very wholesome." The author, known only as "a Lady," also recommended pressing and straining the juice of the cranberry as a drink for the feverish. Mrs. Child's *The American Frugal Housewife* of 1832 introduced cranberry pudding, a batter pudding served with sweet sauce, as well as cranberry pie. Made with nutmeg, cinnamon, and a "great deal" of sweetening, the pie was a logical extension of the long-popular cranberry tart. Mrs. Child also took the opportunity her book of recipes and household hints presented to pass along tidbits of medical lore:

> The Indians have great belief in the efficacy of poultices of stewed cranberries, for the relief of cancers. They apply them fresh and warm every ten or fifteen minutes, night and day.

What the author termed cancers, we would simply call sores. On a topic of less concern—corns—she noted emphatically, "A corn may be extracted from the foot by binding on half a raw cranberry, with the cut side of the fruit upon the foot."

Of course, the early relationship between the cranberry and the turkey continued to be stressed in cookbooks such as H. Pinney's 1848 *Family Receipts,* and the meal suggested had nearly evolved into our traditional Thanksgiving dinner: turkey, cranberry or apple sauce, turnips, squash, and a small Indian pudding.

The cranberries that William Tudor claimed to have seen on every Boston table came not only from Cape Cod, but also from towns west of the city. Cranberries grew in profusion within the meadows along the Concord, Sudbury, and Assabet Rivers, and were harvested by citizens everywhere frontage was found. Wild cranberries joined produce from Middlesex County's market gardens in wagons headed for Boston. Initially a fruit picked mostly for home use, the cranberry became a valuable cash crop in Concord, Wayland, Sudbury, and other towns.

There was one man in Middlesex County who saw the wild cranberry not only as a commodity, but as a thing of value unto itself and watched it with care. Henry David Thoreau kept track of the berry in his journals, where his freshest observations on the world are found. Thoreau knew not one cranberry, but three. Ascending Mount Monadnock, he gathered the mountain cranberry, *Vaccinium vitis-idaea,* stewed it for his breakfast, and proclaimed it "the best berry on the mountain." In Concord, he once made an afternoon of searching Moore's Swamp for the rare, small cranberry, *Vaccinium oxycoccus.* Finding few among the sphagnum moss, he wrote, "There is only enough of these berries for sauce to a botanist's Thanksgiving dinner."

Mostly, Thoreau confined his observations to *Vaccinium macrocarpon,* the American cranberry, which he saw in every season in the meadows. In midsummer, they would set him to thinking of the coming fall: "Green grapes and cranberries remind me of the advancing season." By August's dog days, the fruit had begun to ripen, Thoreau noticed, and recorded it: "Cranberries show red cheeks and some are wholly red, like varnished cherry wood." One September day, friend Sam Barrett showed him some pear-shaped cranberries, different from the common round ones. They impressed Thoreau sufficiently that he sketched one alongside his journal entry: a tiny oblong fruit with stem.

It was Henry Thoreau, the most observant man of his time in Massachusetts, and perhaps America, who also left us our best description of raking wild cranberries, work that took place not only on the rivers he called home, but all across New England. He wrote:

> I find my best way of getting cranberries is to go forth in time of flood, just before the water begins to fall and after strong winds, and choosing the thickest places, let one, with an instrument like a large coarse dung fork, hold down the floating grass and other coarser part of the wreck mixed with [it], while another with a common iron garden rake, rakes them into the boat, there being just enough chaff left to enable you to get them into the boat, yet with little water. When I got them home, I filled a half-bushel basket a quarter full and set it in a tub of water, and stirring the cranberries, the coarser

part of the chaff was held beneath by the berries rising to the top. Then, raising the basket, draining it, and upsetting it into a bread-trough, the main part of the chaff fell uppermost and was cast aside. Then, draining off the water, I jarred the cranberries alternately to this end and then to that of the trough, each time removing the fine chaff—cranberry leaves and bits of grass—which adhered to the bottom, on the principle of gold-washing, except that the gold was that thrown away, and finally I spread and dried and winnowed them.

Like his fellow townsmen, Henry Thoreau probably ate some of the cranberries he collected and sold or bartered the rest. The lowly cranberry may have supplemented the occasional income he received surveying and making lead pencils. But accumulating cranberries brought him little satisfaction when compared to the act of "gathering" the fruit and observing the minutiae of the natural world that appeared even in the flotsam he raked up. Thoreau, as no one else would, saw and noted snowfleas skipping over the cranberries, and amid them, tiny trumpet-shaped cockle shells and caddis worms mixed with meadow grass.

Yet on one occasion, this least entrepreneurial of men considered speculating in cranberries to support a book. In recording the event, Thoreau provided a good description of wholesale marketing at the mid-nineteenth century.

Being put to it to raise the wind to pay for *A Week on the Concord and Merrimack Rivers,* and having occasion to go to New York to peddle some pencils which I had made, as I passed through Boston I went to Quincy Market and inquired the price of cranberries. The dealers took me down cellar, asked if I wanted wet or dry, and showed me them. I gave them to understand that I might want an indefinite quantity. It made a slight sensation among them and for aught I know raised the price of the berry for a time. I visited various New York packets and was told what would be the freight, on deck and in the hold, and one skipper was very anxious for my freight. When I got to New York, I again visited the markets as a purchaser, and the "best of Eastern Cranberries" was offered me by the barrel at a cheaper rate than I could buy them in Boston. I was obliged to manufacture a thousand dollars worth of pencils and slowly dispose of and finally sacrifice them, in order to pay off the assumed debt of a hundred dollars.

No tarts that I have ever eaten at any table possessed such a refreshing, cheering, encouraging acid...as the cranberries I have plucked in the meadows in spring.... Now I can swallow another year of this world without other sauce.

—Henry D. Thoreau,
The Journal of Henry D. Thoreau

For the most part, Henry David Thoreau's involvement with *Vaccinium macrocarpon* took him no further than the river meadows of his hometown. There, with others, he gathered them in autumn. And from that season into the new year, he would eat those berries left behind as he crossed the meadows on his travels. But it was in spring, when the fruits had been frost-bitten and slightly sweetened, that the cranberry meant most to Thoreau. He imbibed and waxed eloquent:

No tarts that I have ever eaten at any table possessed such a refreshing, cheering, encouraging acid that literally put the heart in you and set you on the edge for the world's experiences, bracing the spirit, as the cranberries I have plucked in the meadows in spring. They cut the winter's phlegm and now I can swallow another year of this world without other sauce.

Elsewhere in New England, people took the same pleasure in capturing the wild cranberry as did Henry Thoreau. In places where the berries were especially plentiful and the land commonly held, the gathering was an event organized and regulated by Town Hall. In the *Boston Transcript* of September 26, 1831, the following item appeared:

"Cranberry Day" is hereafter to be a festival at Barnstable. The Journal states that the town authorities had forbidden this valuable berry to be taken from the bogs on Sandy Neck until ripe, and then to pay a stipulated part to the town. September 20 was the day appointed, and 300 men, women and children had a fine frolic. Wet weather has probably reduced the crop one half; but from 150 to 200 bushels were picked.

The *Transcript's* account of Cranberry Day on the common lands that thrust out into Cape Cod Bay could not have been more prosaic when compared to the writing of N. H. Chamberlain. Chamberlain knew this harvest firsthand, and described it with touching fondness in *An Autobiography of a New England Farm House: A Romance of the Cape Cod Lands,* published in 1880. He masked the identity of the towns of Sandwich and Barnstable and collectively called them "Sandowne."

Sandowne had one holiday of its own. It was called "Cranberry Day." Its origin was on this wise. The neck or cape of land that stretched out into the sea, northwardly from Sandowne, had belonged in olden times to the Indians, that is, to nobody, and had in due time, whether in due or undue ways, one cannot say, come to be the common property of the town. On this property, among its sandhills, were almost innumerable patches and snatches of cranberry meadow, where long ago the Indians and the birds had gathered the berries at will. With true Puritan thrift and order, one day had been set apart in the Fall, when every householder in Sandowne, with his household, might gather these berries on the Neck, free of any charge. This was "Cranberry day." It had long since come to be a holiday. It was foretold by the marvelous bakings and brewings of the Sandowne matrons, who at this season, for several days beforehand, were thus accustomed to spend days in preparation for the frolic. It was a day anxiously anticipated for a long while by the younger members of the borough, and had been from time immemorial, so to speak, a day of jollity, frolic, and good-humor. It was a feast-day that occasioned much good-fellowship among the townsfolk, and to the certain knowledge of people curious in such matters, more than several weddings. It was in fact the one pronounced holiday in the Sandowne calendar. …

It was a very pleasant sail to-day on the blue water, for the happy multitude, that shouted to one another from the boats, and trimmed the sails, and laughed and frolicked, and told stories, and cracked their homely jokes, and sailed over to the Neck. It was a pleasant jaunt across the beech-sands to the sheltered spots among the sand-hills, where the crowd settled down among the crimson berries to gather them. It was a busy overseer that Sam Jones made, with his assistants, in allotting to every party their place, and keeping up some sort of order in the picking. It was a happy company that gathered and chatted in the yellow autumn sunlight, a little noisy, maybe, some a trifle rough, but this time intent on frolicking. It was a pleasant dinner that they ate at mid-day, under the bushes; family groups, grandsire, son and grandson, with the goodwife to purvey the Sandowne cookery to her hungry family. Even the dogs, who lay waiting among the beech-grass for their morsel, seemed happy. One remarked how some, who had nursed their wrath a year's time against a neighbor, thawed out to-day under the general good-humor, and chatted again as friends; and how a thousand devices, traits and shades of character displayed themselves in the passing festival in a way very instructing and very salutary to the observer; and the ineffable happiness of the country swains, picking the berries alongside their favorites, or strolling with them among the sand-hills in a silent aimless ramble, that somehow very likely one day brought the two lovers to the parson's was also noticeable.

This was the cranberry harvest painted by Eastman Johnson on the Nantucket moors in the late nineteenth century, a harvest by Yankees who did not pick for a living, but because it was the tradition and one which earned them some extra cash. At remote Nantucket, the scene depicted by Johnson is wonderfully random, not yet subject to conventions. This bog was wild and covered as much by grasses and moss as by cranberries. Pickers are strewn about in clumps and range from tiny children to an aged patriarch who works from a chair. The berries are collected in a wide array of baskets, boxes, and bags. In places where cranberrying was but a minor enterprise, the harvest must have been like this.

Toward the 1880s, Sandowne's "one holiday" was abandoned, and economics soon triumphed over fellowship. It was a fact that the nostalgic Chamberlain bemoaned: "Cranberry day has now for several years been given up as vulgar. The town rents the Neck to pay the taxes. And that is progress!"

***The Cranberry Harvest,* Island of Nantucket, 1880, by Eastman Johnson.**
The Putnam Foundation,
Timken Museum of Art, San Diego, CA

It was a feast-day that occasioned much good-fellowship among the townsfolk, and to the certain knowledge of people curious in such matters, more than several weddings. It was in fact the one pronounced holiday in the Sandowne [Sandwich] calendar.

—N. H. Chamberlain, *New England Farm House: A Romance of the Cape Cod Lands.*

Cranberry vines are still to be found at Sandy Neck and the shore land is again public. Back from dunes, eroded and looking like buttes and mesas, fruit ripen. They share the low hollows with pitch pine, bayberry, and lichen in what looks like a terrarium gone to seed. Any autumn day can be Cranberry Day as the occasional harvester is still seen. Firkins and pillowcases have given way to Tupperware and coffee cans as the picker trades an hour of stooping to avoid the supermarket berry.

Five miles south of Cape Cod lies the island of Martha's Vineyard, and at its west end, the township of Gay Head, known to the Wampanoags as Aquinnah. It is here that the Wampanoag Indians have always lived, making a living from the water, and sometimes, the land. "They annually sell a hundred or two hundred bushels of cranberries, which grow in great plenty in their cranberry bogs." So observed a writer for the Massachusetts Historical Society, who paid a visit to Gay Head in

Gay Head, Martha's Vineyard, Massachusetts.
Lindy Gifford photograph

1807. Picking the wild cranberry there long preceded his commentary. It began, island folklore goes, with the Indian giant Moshup, who dragged his toes in the earth and separated Martha's Vineyard from the mainland. Settling at Aquinnah he plucked whales from the sea for food, his wife and daughters slaughtering them on the Aquinnah cliffs and staining the clay red. They cooked the whale meat over fires of whole trees, torn from a land now barren. In season, Moshup and his people gathered blueberries. But by the shore they picked the greatest delicacy, wild cranberries—a gift from the Great Spirit and a sign, Natives believe, of his care for them.

From Moshup's time forward, wild-cranberry picking on the island would continue. There was a man in Aquinnah whose memory of Cranberry Day extended back to 1905, when he was a young boy. His name was Leonard Vanderhoop, and he was known as Deacon Vanderhoop, or simply Deacon. He took his name from a Dutch West Indian grandfather who came to visit Martha's Vineyard and never left, but he was also a Wampanoag by blood and heritage. From his house, he could see the Gay Head Baptist Church where he worshipped, and the sea, where he worked. A man of natural grace and intellect, for years he was a respected elder in town. The youngest of six, Deacon Vanderhoop followed the common town occupations. He lobstered and worked with other fishermen who ran great wooden weirs, the fish traps that took mackerel, scup, and squiteague in Menemsha Bight. Caretaking for the island's wealthy was also among his trades, and, for the town, Leonard Vanderhoop served as cranberry agent, watching over the bogs until the picking season began.

The land at Aquinnah would not seem to be worth much, a rolling, nearly treeless moor covered by bayberry and bramble, but it used to yield a lot of cranberries in places. The best wild bogs were on the sandy land wedged between the pond and Vineyard Sound at Menemsha. This is what people called "Cranberry Acres." South of Menemsha and across town, folks also picked cranberries below Squibnocket Pond along the beach. But these were not the only places where berries grew. The late Gale Huntington, island historian and a friend of Deacon Vanderhoop, picked each year at the edge of a Vineyard pond. He visited it only on foggy days in autumn, and in this way preserved the secret place from others.

In Aquinnah, Cranberry Day was always the second Tuesday in October. Loaded down with lunches to sustain them, families would head to Menemsha by wagon or ox cart in order to begin picking by nine o'clock. If it had rained the night before or if the dew had been heavy, work was held off until the fruit had dried. Some used what Deacon Vanderhoop called "old-fashioned scoops." These tools, which may have evolved out of the garden rake Henry Thoreau used in Concord, had tiny iron teeth and wooden handles three feet long. One raked the cranberries from a standing position. But there was a more common kind of harvesting.

Of course, then, lots of women here picked by hand. My goodness, they'd get on a place where it was dry—the cranberries would grow up on the sand, you know, on the edges, on the short vines, nice, big, ripe, so dark red it was almost black—and they'd pick by hand. And some of those women could pick an awful lot of berries.

The most important event of the picking came at midday and it came as food and friendship among townspeople. For Leonard Vanderhoop, it was a memory easily recalled.

And then along half past eleven or twelve, they'd quit for their cranberry lunch. And everybody'd get their various ox-teams or horse-teams or whatever. And they'd have a big box of some kind where they'd put their lunch to bring down, and open up their lunch and spread it on a place. Some of them had tablecloths, and put their food out. Pies and cakes and baked chicken.... I guess coffee and tea; they'd make a little blaze with the driftwood. And I think they even had chowders that they would heat, you know, with a little makeshift fire. They'd spend that time with lunch time until about two o'clock. And each one would go around and invite someone to have part of their lunch, if they'd like, you know. It was really quite a gala time.

With a good crop on the bogs, cranberrying would continue for a week or two. Families kept some of the harvest for their own use but sent as many bushels as they could to storekeepers down-island. Vineyard Haven and Oak Bluffs were the towns closest to Aquinnah, so the berries often went there by ox cart. In a very good year, some

Lunch, Cranberry Day, Gay Head, Massachusetts.
Ocean Spray Cranberries, Inc.

Vineyard berries made it to the mainland for sale. What pickers took in return for the fruit were staple foods, provisions that would help them through the winter, when work and money were scarce. The cranberries that remained in Aquinnah homes were much prized, the origin of fall and winter treats such as cranberry pie, cranberry dumplings, and cranberry sauce.

I remember we used to make great bowls of cranberry sauce. These big yellow bowls that stand about this high.... And oh, that was delicious with hot baking powder biscuits and butter and so forth. We had our own cows and cream and the butter we made from it, and oh dear, that was something else. Mmm, delicious. They had more flavor than the cultivated berries.

Aquinnah natives were always careful to distinguish between their wild cranberries and cultivated fruit. To them, the difference between the two was plain, despite the fact that these cranberries were genetic twins.

And the berries were not like the cultivated berries. These were a solid berry. And lots of times, almost as big as a cultivated berry. But the cultivated berries that they grow in the plantations down around—I should know, the place there up near Middleboro. But they grow big, some of them as big as your thumb. And they look nice, taste all right, but they're full of wind.

There is truth in this contention, for a cultivated cranberry is often well fertilized. Fertilizer tends to push the vines into producing more and larger fruit, but it is fruit less hardy, unlikely to keep as well as the wild berry.

As the years went along, some changes came to cranberrying at Aquinnah. The modern cranberry scoop—the straight-bottom and rocker-bottom models popular on Cape Cod—now were used on the island. Eventually, cars replaced the ox carts.

The hurricane, those waves came in from around the Head...like a tidal wave, come right in and broke over the bogs.... Oh, that was something else, to see the salt water kill the vines and everything. Was a long time before they came back.

—Leonard Vanderhoop

During the 1920s, the harvest began to sag somewhat. Commercial scalloping had started on the Vineyard and could make a man far more money than cranberries. Picking shrunk to a day or two for some, then it was fishing. But the event that changed the wild bogs and the traditional picking forever was the 1938 hurricane. On the Vineyard, as elsewhere in New England, it simultaneously inundated and ate the land. Deacon Vanderhoop recalled it this way:

> The hurricane, those waves came in from around the Head—I remember my brother telling me, he was watching—the waves came right in like a tidal wave, come right in and broke over the bogs.... Oh, that was something else, to see the salt water kill the vines and everything. Was a long time before they came back.

They never really did come back. Bayberry and other scrub grew more quickly than new vines and filled up the bogs. The storm changed the way the land drained, too; the bogs were now perpetually wet and produced few berries. At Menemsha, where there had once been acres of wild vines, there are now few. Yet the fall cottage industry of Aquinnah that brought neighbors together for a few days to earn some money and enjoy each other's company is still more than memory, as Deacon Vanderhoop made clear:

> These late years, they always have a commemorative time for the picking of years gone by. Always, every year, they—the tradition—they hold to it and have the younger group go down and pick berries.

Leonard Vanderhoop, Gay Head, Massachusetts.
Lindy Gifford photograph

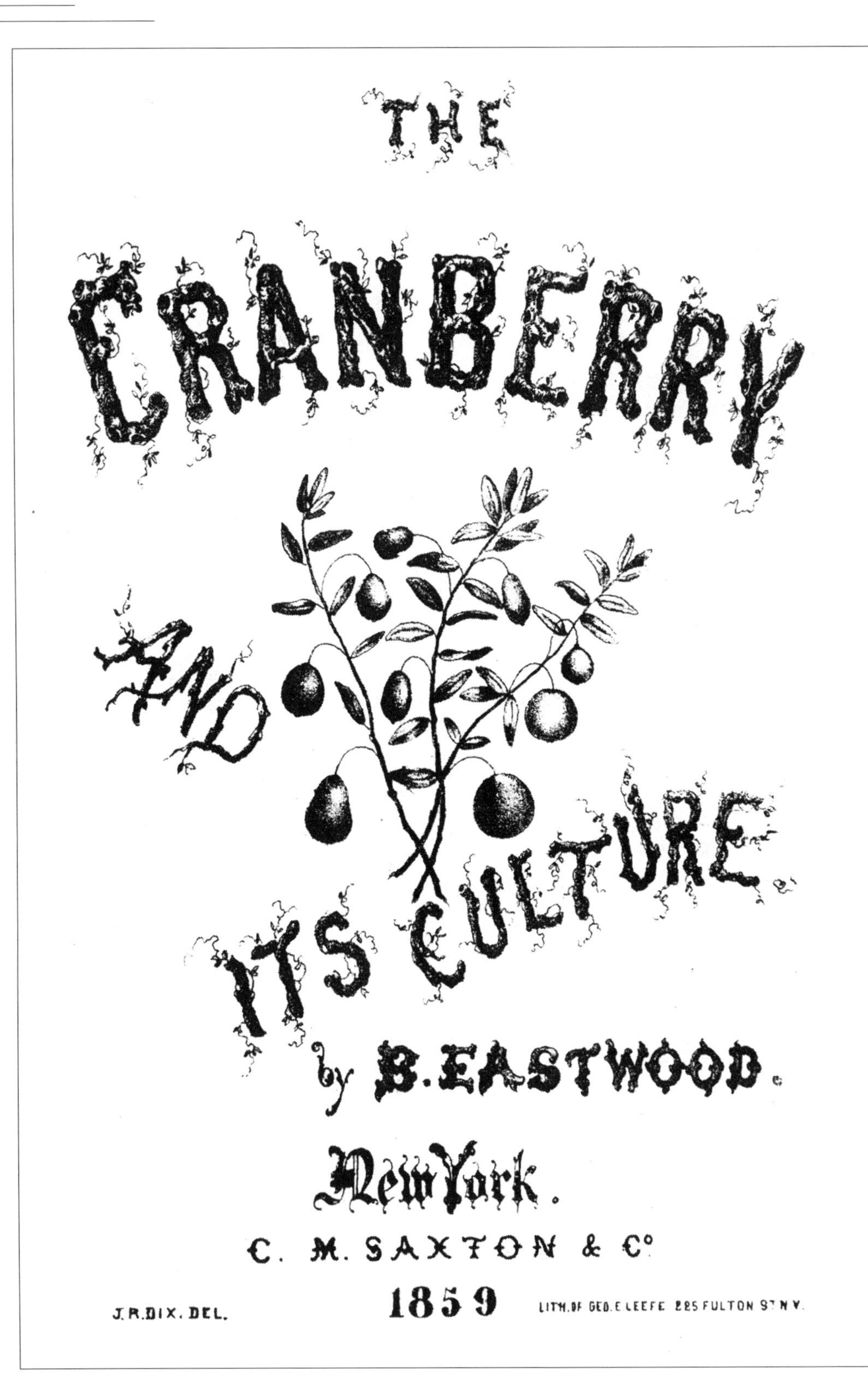

Title page of *The Cranberry and Its Culture* by Benjamin Eastwood.

Chapter 2

Cranberry Fever

He must have been a sagacious man, and bold withal, who saw that he could make the cranberry an article of profitable commerce.

—Rev. Benjamin Eastwood,
The Cranberry and Its Culture, 1859

Until the early years of the nineteenth century, people had been happy enough to let the cranberry go its own way. Come fall, they searched out the wild fruit and picked what nature had left them. It was as simple as that. But at some time between the years 1810 and 1819, the cranberry plant began to fall ever so slightly under the hand of man. That man was Captain Henry Hall of Dennis, midway down the arm of Cape Cod.

Captain Hall's story is the legendary origin of cranberry cultivation. As the tale goes, Hall, a retired mariner, owned land behind the shore and sand dunes of Cape Cod Bay. In places, Hall's land was low, with enough moisture throughout the year to sustain wild cranberries. Over time Hall noticed that the very best cranberry plants on his property grew immersed in sand blown onto them from the nearby dunes. Inside Henry Hall's head, gears meshed and ran, a flywheel spun. He fenced off the vines to protect them from his cattle and watched their progress. The vines spread and bore well, and Henry Hall transplanted other cranberry plants to this productive setting. Once again, the plants took and thrived, doing especially well when the standing water dried away. Hall concluded that sand increased the yield and that swampland became a much better growing environment when drained.

Though ridiculed by many for dallying with a plant whose fruit one could easily pick wild, Hall continued his experiments and observations. For his troubles, Captain Henry Hall was eventually rewarded with a crop that filled thirty hundred-pound barrels and sold readily in New York. In agriculture, observation is the mother of invention. Henry Hall's first bog, the vines mostly run out, can still be found in Dennis. Its framed photograph hangs in the Cranberry Room at the Middleboro Public Library. For cranberry growers, it is an historic place more important than many, including the doubtful Plymouth Rock.

And so the slow process of cultivation began, and for years the process would be inconstant. Into the late 1840s, Henry Hall's routine would be

Wild cranberry vines on sand. Lindy Gifford photograph

carried out in many Cape Cod towns: cultivation was limited to moving cranberry vines from one place to another in a tentative way and waiting to see what resulted. Public opinion did not encourage the experiments. When Dennis's Reverend Benjamin Eastwood wrote his manual, *The Cranberry and Its Culture,* in 1859, he declared, "About fifteen years ago a fair number of people undertook to cultivate the cranberry though they were largely bespattered with unpleasant remarks." Eastwood himself admitted that "the cranberry vine in its wild condition does not seem to offer much temptation to a thrifty farmer, because it appears such a stunted, barren thing, that few would imagine that it could be turned to profitable account." One might as well have tried to cultivate a crop of beach plums.

Many who took up the challenge of cultivating the cranberry worked in total isolation on the sparsely settled Cape. Knowledge of their failures and small successes often got no further than their own houses. What is known of a few early experimenters comes out of the past like brief telegraph messages: The brothers William and Elkanah Sears of East Dennis were tinkering with cranberries not long after Henry Hall. Leonard Lumbert in West Barnstable began transplanting wild cranberry vines in 1836 and continued to do so year after year. Off-Cape, in Middlesex County, known for its wild harvests, Augustus Leland of Sherborn introduced wild vines into some of his wet meadows in 1838. At West Harwich, Isaiah Baker dabbled in cranberry growing during 1840.

But as the years passed, the news of people fooling with cranberries got into print. Local newspapers, which followed agriculture as avidly as modern papers do sports, carried stories of trials and triumphs. Farmers across Massachusetts shared their experiences in journals called *The New England Farmer, The Cultivator,* and the *Annual Report of the Massachusetts Board of Agriculture.* In later years they would be joined by men from all the New England states and the mid-Atlantic who, having read of cranberry culture, decided to try it for themselves. The most notable of these were natives of New Jersey's Pine Barrens, who began transplanting wild cranberry vines in the 1840s and instantly repeated the many trials of cultivation familiar to Massachusetts growers. But persistence paid off, and by the end of the century, cranberry growing was significant in the fragile economy of the Barrens.

In this way, the cultivation of cranberries spread. Unlike the transformation of most other vegetable and fruit crops, it took place so recently and was sufficiently recorded, that the process—the dead ends, disasters, and final victories—can be traced. When scholars went searching for the origins of agriculture in the New World, they found it only in the archaeologists' data. The information came to them as tiny, wizened corn cobs, charred squash seeds, and stone hoe blades. A broad outline of farming's beginnings was there, but not the rich details, the particular and peculiar events of the years. The cranberry allows us this luxury. We can watch its cultivation, the simple encouragement of the wild plant and its domestication, the active selection for certain qualities in the plant species. This is not the stuff of thrillers, but it is always interesting to watch what transpires when man tries to alter a denizen of the natural world.

The first task in taming the cranberry was to create an artificial growing environment as homey as the plant's surroundings in the wild. The soil that plants were transplanted into was not an unimportant matter. In Harwich in 1846, the retired mariner Alvah Cahoon set out vines he had collected in crocus bags into a former laurel swamp covered with loam. The cranberry plants failed to take and died. Cahoon may have thought that if the cranberry could endure peat and sand, it would thrive in the soils most fruits and vegetables preferred. Into the 1860s, novice growers would experiment with varied soil types as Cahoon had and finally come around to what farmer Austin J. Roberts had concluded in 1853. "A boggy or very moist soil," he wrote, "has generally been deemed indispensable to the profitable cultivation of the cranberry, for the reason of its being the natural soil—that wherein nature placed it, and as nature rarely errs, it has been taken for granted that it was not misplaced."

On the other hand, because they had observed the plant in its natural habitat, many Cape Codders were confused about just how much water cranberry vines needed to flourish. Vines had always been collected in freshwater wetlands or from damp hollows along the seashore. If not found growing in water, the plants were at the water's edge. Attuned to nature, the Harwich men Zebina Small and Cyrus Cahoon set out their first vines on swampy land in standing water. They wasted both money and time. What Small and Cahoon did not know is that, while the cranberry vine is a thirsty plant, it will not bear well immersed in water. While "nature rarely errs," it does not always provide optimum conditions for plants and animals. After three failing years, Cyrus Cahoon got the message. He drained his land, which then produced a satisfying crop in the year 1850. Slowly, men were learning how to accommodate the cranberry.

What must have helped these early believers in cranberry culture was a small revolution happening concurrently on Massachusetts farms. In the 1837 *First Report on the Agriculture of Massachusetts,* this news from Middlesex County appeared: "The most remarkable improvements in the county have consisted in the redemption of peat bogs and their conversion from sunken quagmires into the most productive and arable grass lands." The process was begun by cutting ditches across a bog to drain it. Next, the surface was pared and burned, and then it was spread with loam on which grass would be grown. Those living in coastal towns were using the same means to freshen salt marsh and reclaim it for grassland. Also in 1837, a farmer from Essex County wrote to the commissioner of agriculture and declared, "About fifteen years since, I enclosed ten acres by diking. The next season, I drained it with ditches eight inches wide and three feet deep." Those determined to cultivate the cranberry used these newly popular methods—ditching, diking, draining, and turfing—in peat bogs and coastal marshes to create the modified natural environment known as the cranberry bog. For their labors, they got a soil that was not overly wet, yet held water well. Where other farmers spread the pared surface with loam, clever cranberrymen substituted sand, which Henry Hall had found to promote the growth of the fruit. The cranberry bog began to be a discernible physical entity on the landscape: low, vine-covered land surrounded and cut by ditches, sometimes separated from a pond or lake by dikes.

By definition, the peat bogs newly transformed by farmers were not bogs at all, but shrub swamps or wooded swamps in which peat lay. True bogs—interlaced roots of sedges, small trees, and shrubs forming a mat and floating on top of a pond—are rare in the warm climate of southern New England.

Cedar swamp, Carver, Massachusetts.
Lindy Gifford photograph

Cutting and paring turf, from Joseph J. White's *Cranberry Culture*.

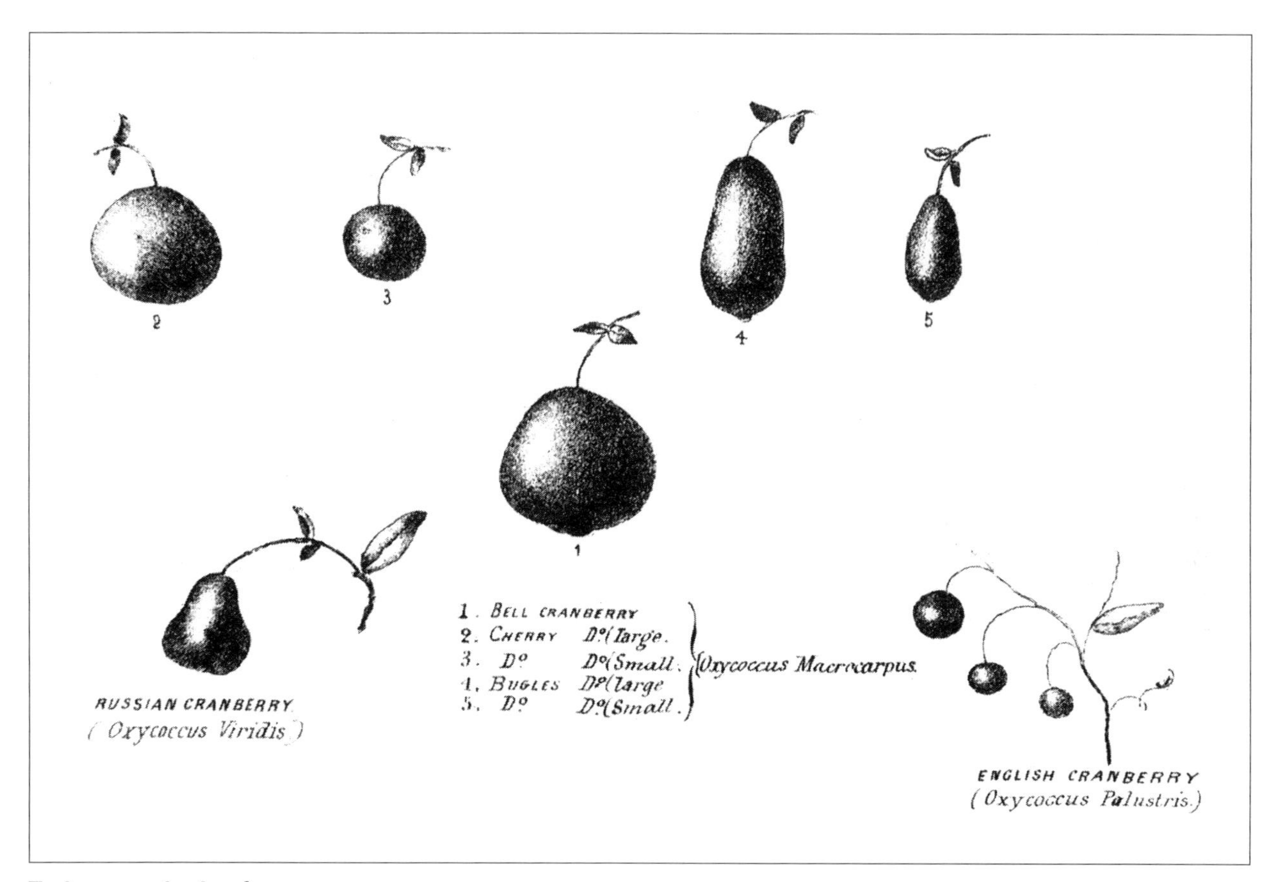

The known cranberries of commerce, from Benjamin Eastwood's *The Cranberry and Its Culture.*

It may prove to be one or more of those instances in which the farmer detects a new species and makes use of the knowledge from year to year in his profession, while the botanist expressly devoted to such investigations has failed to observe it.

—Henry D. Thoreau,
The Journal of Henry D. Thoreau

The delicate bog carpet develops more readily in the moist and cool environment of New England's north, and does especially well in Maine. The swamps that men cleared were old wetlands in the slow process of becoming land, filled up with decayed vegetation and filling in with brush and trees. From the shrub swamps the farmer took out alder and willow, leather leaf and sweetgale bushes. In winter, on ice, he could clear wooded swamps of their far more useful growth: red maple and Atlantic white cedar.

Into the prepared land growers set their vines. The longer cultivation continued, the more selective growers became about the cranberry stock on which they were willing to bet capital and the labor of building a bog. Vines were obtained in several ways. The new grower could dig them out of a wild patch and transfer them to the land he had already made. On Nantucket, Henry Thoreau saw cranberry vines being carted home from swamps, although the overseers of the island's common lands gave notice in the press that trespassers who took away cranberry vines would be prosecuted. Alternatively, one new to cranberry culture could buy his vines from an experienced grower, trusting in his judgment to provide stock that would bear quickly and heavily. Choosing which vines to take from the wild called for an acute sensitivity to slight differences present in the many varieties of the single species *Vaccinium macrocarpon.* Thoreau described the process as it occurred in Concord in 1850:

Horace Hosmer was picking out half a bushel or more of a different and better kind of cranberry as he thought, separating them from the rest. They are very dark red, shaded with lighter, harder and more oblong, somewhat like the fruit of the sweet-briar or a Canada red plum, though I have no common cranberry to compare with them. He says that they grow apart from the others. I must see him about it. It may prove to be one or more of those instances in which the farmer detects a new species and makes use of the knowledge from year to year in his profession, while the botanist expressly devoted to such investigations has failed to observe it.

On Cape Cod, two men chose early and well. In 1843, Eli Howes of Dennis took vines from a place called Basset Swamp. They thrived under cultivation and produced an oblong berry which ripened late—in early October—but was resistant to frost and stored well. In that time, Howes's selection was called the Cherry, but soon it would simply be known as the Howes. Nine years later, Nathaniel Robbins set out vines he had taken from a Harwich swamp. Fellow grower Cyrus Cahoon liked Robbins's choice and promoted it to other growers. The vines had few runners, which made picking easier, and bore early in the fall—around the first and second weeks in September. Named for its shape, it was called the Bell. But as cranberry lore goes, Cyrus Cahoon's wife looked at this berry so deep red as to appear burned, and said, "It's early and it's black." By the name Early Black it is now known.

Other native berries would be chosen by other men throughout the century in Massachusetts: from South Yarmouth, the Bass River; South Carver, the McFarlin and Shaw's Success; Centerville, the Centerville; Plympton, the Whiting Randall; East Sandwich, the Chipman; Marion, the Perry Red; Middleboro, the Middleboro; and Chatham, the Smith—in each name, a reference to the discoverer or place of discovery. But of the more than thirty native varieties known to have been cultivated, the first two, Early Black and Howes—genetically untouched strains—are still the standard early and late cranberries grown in Massachusetts. In an age of relentless plant hybridization, finding fruits whose pedigree belongs only to nature is nothing short of remarkable.

In the 1850s, growing cranberries was as long a shot in agriculture as could be found. The work of draining and clearing several acres of swamp was not inconsiderable, nor was the work of setting vines. Those same vines would not yield a crop for three years, possibly four, and much could go wrong in that time. Winter was especially hard on the bog and always killed patches of vine, which showed up bleached and blighted in the spring. When the vines blossomed early in the year, and again when they bore fruit in the fall, chances were good that there would be a frost or two. Like the peas planted too early and the tomatoes set out late, the cranberry was not spared. The frost killed the buds instantly and softened the berries, making them difficult to sell.

To increase the chances of gathering a crop, the cranberry fruit and vine had to survive damaging weather. In the 1840s in Osterville, Jarves Lovell sheltered his vines against frost with cotton cloth suspended several feet above the bog. Others blanketed their bogs with eel grass gathered at the shore when a cold snap was pending. But the method of protection that would prevail came again from observing the cranberry in its natural environment. It may have happened for many growers as it did for Thoreau, who, after returning from a row on the Sudbury River, noted in his journal, "I see many cranberries on the bottom, making a great show. It might be worth the while, where possible, to flood the cranberry meadow as soon as they are ripe and before the frosts, and so preserve them plump and sound till spring." Thoreau's notion that water protected both berry and vine from frost and winter was accurate and one that others were coming around to. In 1853, the year that Henry Thoreau had his revelation, Augustus Leland of Middlesex County also suggested that flowing or flooding the vines combated frost. For Leland, water served another purpose on the bog: he used it from spring into summer to drown the cranberry worm, which infested his acres yearly.

Yet the grower who could take full advantage of water as a deterrent to frost was rare, for it involved owning land with a body of water on it. If a stream ran through the property, it could be channeled through the bog, or the bog built around it. To use the water, growers borrowed the water control techniques that had arrived in America with English settlers of the seventeenth century. With gates installed in the dikes at either end of the bog, the farmer had some insurance against the weather. The stream above the diked bog formed a pond, a ready source of water with which to flood the bog if a frost was imminent. And throughout the year, the grower could allow just enough water to flow over the upper gate, around the ditches, and out the lower gate to keep the pond full and the bog irrigated. In winter, the lower gate could be closed to again flood the bog and protect the vines from wind and cold. Farsighted cranberry growers tried to build a number of bogs along the course of a stream. As if to suggest that their use of an old technology was somehow novel, cranberrymen changed the names of its features. What millers knew as dams were called dikes by growers, whether they held back a pond or just separated one piece of bog from another. Water to power the miller's wheel passed through a flume and was held back by a flume gate; cranberry growers referred to their gates simply as flumes.

Many cranberrymen had to depend on the weather for whatever water their bog received. Natural hollows among seashore dunes had been prized bog locations since Henry Hall's success there. The vines depended solely on the water table and rainfall for moisture in these places; deliberate flooding was impossible. The best that could be hoped for was enough winter rain to cover the vines. The same was true in coastal salt meadows, where men diked off a piece of land, freshened it, and planted. If the plants could make it through winter's cold and wind and then spring frosts, the bog was well protected come fall because, the sea, which gradually stored the heat of long summer days, radiated heat to the far colder air, keeping frost at bay. The drawback of these coastal bogs was the threat of ocean storms, which could inundate the land with salt water, killing the cranberry vines. Bogs that could not be flooded—whether in a dune landscape, at salt marshes, or in former wetlands—were known as dry bogs, and in the nineteenth century it was far more common for a cranberry bog to exist without benefit of water than with it.

Dry cranberry bog, the Carleton Farm, East Sandwich, Massachusetts.
Courtesy Carleton family

It was in 1855 that officialdom first took notice of the cranberry and included it in the agricultural census issued by the secretary of the Massachusetts Board of Agriculture. The figures listing acres of cranberries and their value by county were nestled in between the statistics on sheep and wool and beeswax. The value of the Massachusetts cranberry crop was placed at $135,000, making the cranberry more valuable than such staples as poultry, eggs, turnips, maple sugar, and pears. What was printed is surprising, for Middlesex County was acknowledged to be the state's leading producer, with 2,554.375 acres of berries, followed by adjoining Norfolk County's 897 acres and Worcester County with

Cultivated cranberry vines are those upon which only the labor of transplanting from the fresh meadow has been bestowed.

—farmer John Howard,
Annual Report of the Secretary of the Massachusetts Board of Agriculture

641.25 acres. Next came a second tier of cranberry-growing counties: Bristol, along the Rhode Island border, with 380 acres; Essex, just south of the New Hampshire line, with 370 acres; and Plymouth County with 361.5 acres. Only then appeared Barnstable County—Cape Cod, the birthplace of cranberry culture—with 197 acres. Trailing along behind the Cape were the islands: the smaller Nantucket with 19.75 acres of cranberries and the far larger Martha's Vineyard where 14 acres of fruit grew. The central and western Massachusetts counties of Hampden, Franklin, and Hampshire grew a combined total of 27.375 acres. Only Suffolk County—Boston proper—and the western county of Berkshire failed to produce a crop at all.

That Middlesex should have led the counties in cranberry acreage is not peculiar, since both natural and cultivated cranberry bogs were counted in the census. It is estimated that between eight and ten thousand acres of meadowland once bordered the Concord and Sudbury Rivers, which wind through the towns of Wayland, Sudbury, Lincoln, Concord, Bedford, Carlisle, and Billerica. It is not inconceivable to imagine several thousand acres of river meadow containing cranberry vines. The conditions that gave Middlesex County such vast cranberry acreage also prevailed in the five other Massachusetts counties with greater acreage than Barnstable. Their geologic history had produced rivers or streams bordered by wet meadows and vast swamplands, all perfect habitat for wild cranberries. Cape Cod's geologic past had been different, leaving a landscape noted for its hundreds of discrete kettle-hole ponds, less perfect environments for wild cranberries. In the cranberry acreage recorded for each county, the great majority was comprised of wild vines, the minority by wild vines nurtured by curious men. The Massachusetts Board of Agriculture's decision to combine the acreage of natural and cultivated bogs in the census points out how small a difference was perceived between the two. "Cultivated cranberry vines," wrote farmer John Howard at the time, "are those upon which only the labor of transplanting from the fresh meadow has been bestowed." For the most part, he was right.

The image of Cape towns as the heart of this fledgling agriculture was implanted because it was here that cranberry growing thrived and became a way of life. The cranberry industry had become so well established on Cape Cod by the 1880s that historians were already looking back to the early part of the century to discover who its pioneers had been. Had cultivating not soon declined in Essex and Worcester Counties, we might know more about the early experimenters who lived there. Innovators in cranberry culture were, in fact, found throughout the state, and contributed, known or unknown, to what became the established methodology of growing. They were at work before 1850 in Danvers, Winchester, and Essex, where perhaps they took their vines from the great expanse of beach the town shared with neighboring Ipswich. By this time, Addison Flint had built a flume at his bog in North Reading, and Libbeus Smith of East Bridgewater had set out some cuttings he had received from a grower in Bellingham. And since 1840, Augustus Leland had been battling the cranberry worm at his Sherborn bog.

In 1859 the cranberry industry got its first handbook on growing, a sign that it had indeed arrived. *The Cranberry and Its Culture* was the work of a minister from Dennis, the Reverend Benjamin Eastwood, and grew out of the tremendous response from people everywhere to some columns on cultivation he had published in the *New York Tribune.* It seems appropriate that a preacher should have been called to write on the cranberry, for growing was being taken up with a sort of evangelical fervor.

"Work of the Fruit Worm" from *The Cranberry and Its Culture.*

Eastwood's treatise was simple, upbeat, and encouraging, like a good sermon. Its message was this: Be a thrifty American; turn your swampland to profit! With much speculative information about *Vaccinium macrocarpon* around, Eastwood tried to lead his readers down the path of righteousness. He told them where to plant: pond edges, swamps, pond bottoms, meadowland, and low, sandy patches were all fine. He admonished them to avoid at their peril the "upland mania," for some growers were attempting cultivation in loam or clay soils on land that was high and dry. Eastwood's warning was timely, for unscrupulous nurserymen were already selling vines supposedly suitable for upland situations. Such conditions led to blighted vines with few berries, or plants scorched and burned up altogether. The Reverend Eastwood reminded his flock that the cranberry needed water, but not too much and not too little. He explained that one could gather vines for planting in spring or fall, as they wintered over well in a cellar. In May or June, vines were set out in sods, as single-rooted vines or as cuttings. It was even possible to plant cranberry seeds, though the results were less encouraging. What resulted was, in Eastwood's parlance, a cranberry "yard," not a bog. The name probably derived from the proximity of the plantings to the grower's home and grounds.

About some things, Reverend Eastwood was overly sanguine. Weeds, for instance, were a problem he considered easily overcome. In the first few years before the vines filled in, the cranberry grower could hoe around his plants like any other farmer. Then came Eastwood's speculation on the hereafter: "But generally after the second or third year's careful cultivation, the vines will take care of themselves and will eat out weeds and grass, and thus leave little to be done by the grower." Grasses, sedges, shrubs, and brambles have long outlived this prediction. Of insect pests, Eastwood admitted, "There is the worm. We have not seen it and have only met with one grower who has." The cranberry worm, who makes his living eating cranberries, would soon be on intimate terms with every grower. In some years, he would claim one-third of the Cape Cod crop.

As it was, many New England farmers went their own way, ignoring agricultural books and journals for homegrown methods. Some were long familiar with the cranberry or fruit worm, and had taken steps against him. James Lovell had fought the worm since 1844 and did so by spreading wood ash and lime on the bog. Augustus Leland flooded his bog until the first of July each year to drown the worms. The noted Professor Agassiz of Harvard devised a more devious method, suggesting that fires be built around the bog to attract and destroy the fruit worm in the miller stage. The advice of Reverend Benjamin Eastwood, both good and bad, was likely never read by every cranberry grower.

"Cranberry Fever" was the phrase used by newspaper editors to describe the acceleration of Cape Cod's economic pulse that cranberry growing brought there in the late 1850s. It was a boom that would not be matched on Massachusetts's narrow arm until great waves of tourists began to arrive at the close of World War II. In 1853 the secretary of the Massachusetts Board of Agriculture estimated that the cost of preparing an acre of bog was about three hundred dollars. Barring serious frosts, droughts, or insect infestations—greater threats than many eager investors realized—one could hope to harvest a bushel to a square rod, about 150 bushels per acre, within three or four years. At the 1853 price of two to four dollars per bushel of cranberries, an acre of bog could pay for itself in three years; from then on it was mostly profit. Farmer Addison Flint of North Reading spoke for growers on and off the Cape when he wrote:

> I have no doubt but there is swampland enough in Massachusetts suitable for raising cranberries, to raise enough, at the prices they brought for the last two years, to come to more than all the corn, grain, and apples raised in Massachusetts. I will here add the remarks of two gentlemen in regards to glutting the market with the article; the first a city man, who said, the inhabitants of Boston and New York have not yet begun to get the taste of cranberries. The second, a farmer and nurseryman, who said, if I had ten acres and every man between you and me, and every man between you and the Canada line had ten acres each, and they all bore two hundred bushels the acre, it would not glut the market.

I have no doubt but there is swampland enough in Massachusetts suitable for raising cranberries, to raise enough, at the prices they brought for the last two years, to come to more than all the corn, grain, and apples raised in Massachusetts.

—farmer Addison Flint,
Annual Report of the Secretary of the Massachusetts Board of Agriculture

---

Every man with a little swampland and the smallest amount of business acumen made that land over into bog. What had formerly been worthless now commanded a price of $1,000 per acre, at the least. "Individuals have received as much as $2,000 and $3,000. The epidemic is spreading to all classes," wrote a Cape Cod newspaperman. The *Yarmouth Register* observed, "Laborers find welcome employment at good prices and 'our capitalists' are learning their money can be invested to better advantage at home than in the West." Cranberries were even referred to as red gold. In reality, the vagaries of cultivation were such that one's chances of striking it rich were not much better on a Harwich bog than in the foothills of the Sierra Nevadas.

Despite the risk, it is no surprise that so many Cape Codders gambled on cranberry growing. Cranberries joined the very limited resources of Barnstable County for which there was a market elsewhere. First and foremost, the Cape had always been a fishery; its fleet caught almost all the codfish and mackerel consumed in the nation. Salt fish was coasted in schooners to Boston, together with sea salt evaporated out at coastal salt works. The only other notable product of the region's sandy soil was onions. Cape Cod's exports were the basis of a good chowder, but hardly a thriving economy. Packed in hundred-pound barrels after picking, drying, and a good winnowing, cranberries also traveled to Boston by water. Once arrived, they were handled by early commission merchants such as Curtis and Hall, who either sold them to shopkeepers or reshipped the berries to New York and other eastern cities. In time, schooners would sail cranberries to the West Coast, with San Francisco the principal market.

The cranberry must not have remained in any market long, for the demand was immense and seemed insatiable. Ever since William Tudor had noted its fashionableness early in the century, the cranberry had become more in vogue as cultivation made it more available. "The wealthy would as soon part with the apple as with the cranberry, and it is the rage among the rich," noted a Cape Cod observer. It was relished equally with goose or mutton, and was good—not to mention stylish—as pie or in a pudding. Craved in the cities, the cranberry's presence in its own home was transitory. "In the immediate neighborhoods where cranberries are cultivated," the observer continued, "few are consumed."

The middle years of the nineteenth century brought forth a number of glamour crops in New England, of which the cranberry was but one. In the Connecticut River valley, a newly cultivated tobacco soon known as Connecticut broadleaf caused a similar pattern of excitement and speculation, and in some years, brought huge profits. Like the cranberry, tobacco would remain an agricultural mainstay of its region. Another crop proved to be a mere flash in the pan, running its course in less than twenty-five years. In every New England state, men planted mulberry trees and introduced silkworms to feed on the leaves. The result was cocoons from which silk was spun, principally in Connecticut and Massachusetts. Silk was sold for cash or exchanged in barter by farmers, until hard weather and mulberry blight killed the industry around 1840. Other crops, unknown today, rose and fell. Broomcorn for brooms and brushes flourished in western Massachusetts, and hops proved profitable for cultivators in Middlesex County. In the end, only a few crops were favored by optimum growing conditions and the constant demand necessary for a long life. Fortunately for many, the Cape Cod cranberry was such a crop.

The sand plain, Plymouth County, Massachusetts.
Lindy Gifford photograph

There can be no doubt that cranberry culture had taken hold and flourished on Cape Cod. When the Massachusetts Board of Agriculture released its census figures for 1865, Barnstable County had jumped to first place with 1,075 acres of cranberries. The figure did not represent the exploitation of the Cape's limited natural cranberry habitat, but the building of 878 acres of bog. That amount of acreage was more than Middlesex County now claimed, due to the drowning of its river meadows by damming. Bog acreage also shrank in Norfolk, Worcester, Bristol, and Essex Counties, perhaps for the same reason. Yet cranberry acreage increased in Cape Cod's neighbor, Plymouth County. Men were already responding to the call of the Board of Agriculture, which wrote in 1863, "Places favorable to the culture of this fruit are to be seen throughout the county, and it is to be hoped that the owners will awake to their interests and commence the culture without delay."

Cranberry-bog acreage would continue to increase on Cape Cod and in Plymouth County as it fell away over the rest of Massachusetts and New England. The prominence of these regions would never wane, due to an advantage that growers elsewhere, for the most part, did without. That asset was sand, the bountiful gift from New England's last great ice age of 15,000 years ago. The land left in the glacier's wake was at first nearly as favored as any other. A good humus layer had built up over the centuries, and beech, maple, elm, cedar, and oak grew there. But settlement took its toll, especially on Cape Cod, and forests became ships, houses, and firewood. Forest fires were not uncommon in these woodlands, badly dried out by the constant southwest winds. By 1802 firewood for the town of Chatham came from Maine. As the protective cover of the forests disappeared, the Cape began to lose its good soil to erosion. In the 1840s and 1850s, thousands of white and pitch pines were frantically set out to hold the earth. What resulted on Cape Cod and up into Plymouth County is known to botanists as the sand plain community, a level land of porous, sandy soils studded with scrubby bear oaks and pitch pines.

The fruitfulness of cranberry vines when covered with sand was one of Henry Hall's first observations. Successful growers made a practice of spreading a thin layer of sand over their bogs after the harvest, either on winter ice or on the vines themselves in early spring. Sand works in several ways to encourage the vines: it discourages weeds and moss, makes a good mulch, radiates heat at night to ward off frost, and drains better than the underlying peat, making it a good growing matrix. Without a good dose of gravelly sand every three or four years, a cranberry bog's yield will eventually dwindle to nothing. Growers outside the two counties could get sand, but it was rarely close at hand. For a price, it could be bought and carried to the bog by horse cart. In Plymouth County and on Cape Cod sand was everywhere, as close to the grower as the edge of his bog.

Cranberry shipping box and barrel.
Lindy Gifford photograph

Chapter 3

A Way of Life

When some promising industry, some added means of making a living comes to a place and thrives, it is quickly integrated into life and becomes an established part of the economy and local culture. In the latter years of the nineteenth century, the 1870s and beyond, this was the role cranberry culture played on Cape Cod. The slender economy of the Cape, long dependent on fishing and coastwise trade, was considerably fattened by the cranberry, which brought new money to pocketbooks in every town. Growing was, for the most part, small-time. Families had an acre or less to which they gave some care over the years. Cultivating the fruit continued to be a very fickle thing; insects and frost had as much to do with the year's crop as the grower did. In few cases were the proceeds from the bog anyone's sole source of income. Some had a profession and a bog on the side, others a string of business interests or part-time jobs of which cranberry growing was but one. This was the experience of Albert Ryder from the village of Cotuit, as his son Malcolm recalled in the kitchen of his father's house.

My father and his brother, Uncle Wallace Ryder, they inherited a painting business from his father, my grandfather, who was Joshua Ryder. The paint shop and business was right here on the place. The shop I tore down, 'twas a two-story building. The upstairs, he painted carriages and skids. Came down and hauled them up with a tackle. They had the varnish room that was a plastered room and was tight, the floors were wet down. As a small boy I used to go up and watch the procedures. And then my father, he also bought a painting business in Falmouth that he carried on for about fifteen years. And then he had a very active oyster business in Cotuit with Uncle Wallace Ryder, they carried on for about fifteen years. In the meantime, he accumulated different bog properties.

The cranberry also took its place among the crops harvested on the farm, the most common economic unit in nineteenth-century America. Such a one was the 250-acre Carleton Farm in East Sandwich, bordering tidal Scorton Creek and its lovely marshes. John Carleton's minister father promised him a farm on his graduation from Harvard College, and in 1891 Reverend Carleton made good on that promise. The farm had a dairy, chickens in poultry houses, and good ground for vegetable crops. Fruit of every sort was grown by the Carletons: Baldwin, Gravenstein, and Russet apples

The Carleton Farm, East Sandwich, Massachusetts.
Courtesy Carleton family

The berries when gathered showed signs of blight, and were unsalable. Thousands of bushels had to be thrown away.... This made a great hole in living funds and as many here have had to mortgage their homes in order to eke out the family funds, many are now left high and dry for ready cash. Trade stops, the stores are hard up, and all feel the hard times.

—William Morse, Barnstable

in two orchards; two kinds of raspberries; cultivated blackberries; Concord grapes; and cranberries. John Carleton's cranberry properties—called Big Brown Swamp, Acre Bog, Sally Jones, and Reservoir—were smallish dry bogs. Several were set behind the house and buildings, between the farmstead and the great stretch of beach known as Sandy Neck. In these early years, cranberry growing was of such minor importance to John Carleton that he packed the crop in a spare room of his house. Not until the next century would cranberries become the premier cash crop on the Carleton Farm.

In a good year, a cranberry bog paid back part of the cost of its construction. If it was owned free and clear, it might pay the taxes or provide some amenities for the family. But the economic return of cranberry bogs varied greatly, depending on the bog itself, the care given it, and the vagaries of nature. In the rush to make cranberry money, many built bogs on poor soil or on land too low and wet or high and dry. Bog was wedged into every conceivable place that offered the slightest chance of success. Cape men continued the traditions of the earliest cultivators and built many dry bogs: pieces that could not be flowed from a pond or stream and depended on rain for moisture and frost protection. Dry bogs bore perfectly well, except in years when drought struck or frosts came early. Yet much cranberry bog—marginal for whatever reason—offered little return on investment and dropped out of production.

What people read or heard about were success stories like those related in James Webb's book *Cape Cod Cranberries,* published in 1886. Webb's case histories consisted solely of bogs that quickly provided a stunning cash return. The most notable was the bog of the Newtown Company. Built between 1864 and 1866, the property was sixteen acres in size and constructed at a cost of $6,800. Over the next fifteen years, the bog paid dividends of more than $45,000. There were other investments made that the public did not know about, however. These were hidden away in account books like those of cranberry grower and schooner owner Archelaus Baker. Baker built a small bog in 1893 costing $161. Thirteen years later, he had gotten back from it only the cost of construction. Individual tales of a profit soon made or an investment that languished could seesaw back and forth forever. When Simeon Deyo, author of the substantial Victorian tome *History of Barnstable County,* wanted to know how profitable cranberry bog investments in his time were, he went to the pioneering grower and promoter of Early Black vines, Cyrus Cahoon of Pleasant Lake, Harwich. He wrote that Cahoon "fairly expresses the belief that the total investments in this industry in Barnstable County since 1850 have yielded an average annual return of thirty percent, although this average includes some recent years wherein some growers have made total failures."

Those seasons of total washout sometimes found a sensitive recorder who saw the effects on people in small towns and noted the growing dependence of Cape Codders on the cranberry. One was William Morse of Barnstable:

> April 17, 1897, called on Asa Bearse at his store and in the evening Roland Nickerson. Both had doleful tales of the slow hard winter following an unfruitful autumn, for a severe hailstorm, they said, fell on the cranberry vines in October, beating down the berries before they could be picked. The berries when gathered showed signs of blight, and were unsalable. Thousands of bushels had to be thrown away. Twenty-five thousand dollars [were] lost.... This made a great hole in living funds and as many here have had to mortgage their homes in order to eke out the family funds, many are now left high and dry for ready cash. Trade stops, the stores are hard up, and all feel the hard times.

Cranberry growers and promoters had tried to ensure that if the wolf was at the door, money lost on a cranberry bog was not the cause. In the 1860s the industry began to expand beyond family ownership of small bogs. Experienced growers looked to large properties—fifteen or twenty acres—as a means to greater profits. The problem they faced was the financial risk. The cost of building a sizable piece could be as much as ten thousand dollars, and if the enterprise failed, an individual could be crushed. So men turned to a mutual insurance scheme known on Cape Cod for many years, and sold shares in bogs to finance their construction.

The sale of shares—sixteenths, thirty-seconds, and sixty-fourths—was the common means of building and operating the coasting and fishing vessels that were the Cape's economic mainstay. If an investor owned a thirty-second of a ship that was lost at sea, his loss was not total, and his interest in other boats might cover it. Many of those first willing to chance growing the cranberry had followed the sea and held shares in vessels. When success with cultivation encouraged them to build big bogs, they chose Cape Cod's standard means of raising capital. In some cases the investors were members of the same family, but in others, they were men and women unknown to one another, simply looking for a good return on their money. The system worked well only in the short term, according to the late William E. Crowell, South Dennis cranberry grower and retired lawyer. Mr. Crowell knew, for he both managed and worked the bogs his grandfather had built on shares in the nineteenth century.

During the cranberry season, the gathering of the crop is paramount to everything. Schools are closed, or poorly attended, houses stand tenantless. No travel is seen along the quiet highways during business hours. But should a stranger meet the early morning motley crowd, lively, good-natured, independent, in the heavy, slow work-wagons, going to their day's work, he could not help seeing that they mean business. The Cape Codder does not work because he must. By no means. The thought is an insult.

—O. M. Holmes,
"Cape Cod Cranberry Methods"

When the people started in who owned the shares, they were usually prominent business people, people who had money and so forth and invested in bogs. But over a period of years, the ownership descended and it was in the hands of widows and minor children and that sort of thing, who had no way of contributing back to the bog itself.... And it was a very good idea so long as the bogs are new and all producing, and so forth, but when they got old and some of them didn't produce so well, it was an embarrassing situation.

Years later consolidation of ownership occurred on these bogs that began with so many owners, making for some peculiar situations. It meant that Malcolm Ryder's father was forced to settle with eleven different owners in 1905 to buy the Abigail Brook Bog in South Mashpee. And although the Ryder family bought half interest in a property known as the Hannah Bog in 1895, it was not until 1961 that Malcolm could purchase the remaining half. Remnants of the share system long persisted. In Osterville during the 1980s, a retired Cape Verdean man worked three-fourths of an acre left to him by a former employer. It sat, however, right in the middle of fifteen acres of bog held by another grower. To the great consternation of this fellow, the owner of the small piece planned to hold his ground until the day he died.

Late in the 1800s, one part of cranberry growing became formalized beyond all others: the harvest. By that time, the culmination of the growing year had evolved its own standard means of practice, a sort of etiquette. Today the harvest of the first berries, Early Blacks, begins about the fifteenth of September, but in the 1880s picking might begin just after Labor Day. Gathering berries by hand was a slow business; to assure getting the

Hand picking, Harwich, Massachusetts.
Harwich Historical Society

crop in before a frost, the first fruit was picked a bit green to redden in the box. As the harvest continued, the cranberries ripened.

If there was bog in your family there was no question of whom you picked for. But should your family do other work, your opportunities were broadened considerably. The harvesters were all manner of Cape folk willing to spend Indian Summer working outdoors: youngsters, young adults, men, and women. Since many growers paid by piecework, pickers who were free agents would "prospect" the bogs in town, as picking season grew near. By harvest, they knew which bogs had the heaviest crop and to these places they quickly went to make good wages. A bog with a thin crop was harvested last of all. There were also picking contractors on the Cape, the sole professional element in the fairly simple, unhurried work. On larger bogs, the contractor would agree to pick, pack, and transport to the rail station another's crop for an established price per barrel. To do so, he hired harvesters and screeners and transported them to the bog. If there was a bog house on the property, the crew would camp out until the job was completed. The day was given over to work, the night to a little visiting and fun. The butcher and grocer made frequent stops. For the duration of the picking, the cranberry bog was a self-contained world.

On Cape Cod, autumn was about nothing so much as the harvest. It appeared, just as O. M. Holmes described it to the New Jersey growers in 1883.

During the cranberry season, the gathering of the crop is paramount to everything. Schools are closed, or poorly attended, houses stand tenantless. No travel is seen along the quiet highways during business hours. But should a stranger meet the early morning motley crowd, lively, good-natured, independent, in the heavy, slow work-wagons, going to their day's work, he could not help seeing that they mean business. The Cape Codder does not work because he must. By no means. The thought is an insult.

In the reminiscences that remain, the cranberry season comes across as an idyll, a month of work and fellowship among small-town neighbors. They joked and flirted, shared their lunches, and earned money for school clothes or Christmas presents. It was largely a Yankee harvest. The feeling was encapsulated by a Yarmouth grower writing to an off-Cape friend in 1891: "We are having fine weather and a good time picking berries, and when we get the decks cleared up we will have a Jubilee."

The business of hand picking was almost unbelievably labor intensive. Consider Malcolm Ryder's account of a story his father told:

To go back to the Well Swamp that he built in South Mashpee, when that first came into bearing, he told me many times that with the families that were there, there were about a hundred on that bog to pick it by hand. That was only a matter of about fourteen to fifteen acres. Handpicking was a slow procedure.

The day never began early, because it could not. If the berries were picked still covered with dew, they rotted in the barrels, so harvesters waited, sometimes happily, sometimes anxiously, for that moment falling between nine and ten in the morning when all was dry and ready. If rain had come the previous evening or a frost had compelled the grower to flood the bog, the day's picking could be lost entirely. Once on the bog, what the pickers found was often the same, the acreage lined off with string to form alleys that held as few as one or as many as four workers. Friends tried hard to be paired in the same alley, as did young couples; conversation kept up the spirit and relieved boredom. Most were similarly armed: a tin bucket—known as a six-quart measure—to pick into, a lunch pail with substantial contents for the nooning, and finger stalls, armlets, and hats of straw or cloth. Stalls shielded the fingers and armlets the wrists, from the tough and raspy vines that otherwise broke the skin and rubbed it raw. Protection from the vines also dictated the bog wardrobe. "It is a sight which must be seen to be appreciated," James Webb wrote, "as each individual is dressed in a costume of startling originality and of the most unique description." Human scarecrows were at work.

The rules of the harvest were these: Each picker's name was registered with the tally keeper or "tally," and a number assigned. The picker, in turn, attached a tag with that number on the tin measure. Work then began and when the measure was "crowning full," the picker called out the number to the tally and then dumped the measure into a bushel box. The tally kept a running score of each worker's labors for the length of the harvest.

Within the alleys, the pickers moved along on their knees gathering berries in a posture once described as "being doubled up like a jackknife." It was always well to keep apace of the others in your alley or they might be tempted to "scrouge." Scrouging was the Cape term for trespassing into another's line of picking when the fruit was particularly heavy there. Scrougers never got too far, since an overseer—either the grower or his foreman—was always on the bog to keep order among the pickers. The overseer's prime task was to make sure that picking was clean. It was easy to gather the tops, the cranberries that stood out on the vines;

Tallying.
Ocean Spray Cranberries, Inc.

When that [bog] first came into bearing, he told me many times that with the families that were there, there were about a hundred on that bog to pick it by hand. That was only a matter of about fourteen to fifteen acres. Handpicking was a slow procedure.

—Malcolm Ryder

The overseer.
Harwich Historical Society

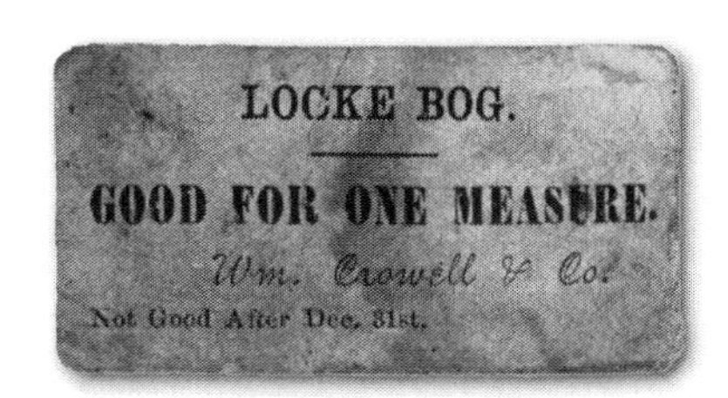

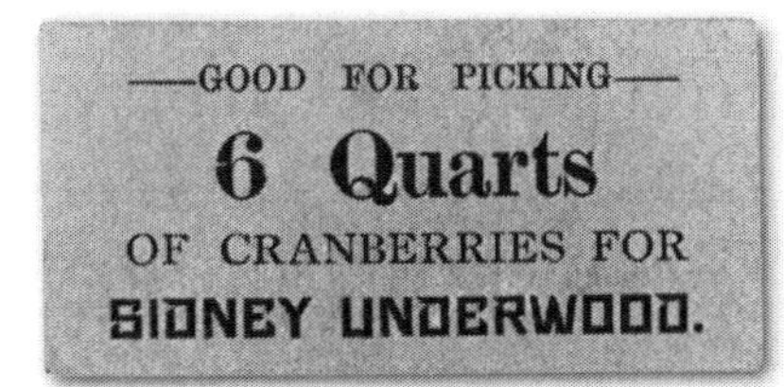

Measure tickets. Harwich Historical Society

the overseer ensured that the bottom berries, growing close to the bog and reached only by a hand thrust down into the prickly vines, were also taken. Even at that, it is estimated that hand picking left about 15 percent of the fruit on the vines, a substantial loss if a hundred-pound barrel brought ten dollars to the grower.

For their labors, pickers got about ten cents for every six-quart measure, and, on the average, picked about a bushel and a half of cranberries in a day. But money was not forthcoming until the harvest's end. Small paper tickets, "Good for One Measure," were the currency given out daily to the workers. These were either exchanged at the close of the season for cash or used at local stores during the harvest. In a month's time, a picker neither fast nor slow would make twenty-five dollars, very good wages for the time.

Once in field boxes, the cranberries were wheeled off the bog by workmen with handbarrows for sorting and packing. Before the early 1900s, when the cranberry separator became popular, the work of sorting must have been both laborious and nearly blinding. The equipment involved was simple enough so that the work could be done at the bog edge or in a building where the crop was stored. The essential item was called the screen, a flat-bottomed trough about six feet long. It measured a yard or more across at one end and tapered to the width of a barrel top at the other. The screen's name derived from its bottom, a series of pine slats a quarter of an inch apart. Propped up on barrels, the screen was worked by a few women who dumped crates of berries in at the wide end, and worked them over, small berries, twigs, and leaves falling through the slats. Screening meant running your hands over the berries to feel for soft ones, or picking the bad fruit out by sight, as you pushed thousands by. It was work that required a good sense of touch and good light.

The cranberries came to rest in hundred-pound barrels supplied by coopers who worked in every Cape Cod village, shaving the oak staves and forging the iron hoops themselves. Barrels were the standard containers for many perishables in the 1800s, but even so, the cranberry industry provided a major new outlet for barrel makers. With the barrel filled tightly and the head pressed in place with a large hand screw, the fruit was ready for the market.

Screening cranberries.
Harwich Historical Society

Heading barrels.
Ocean Spray Cranberries, Inc.

The majority of the Cape Cod cranberry crop found its way to Boston and New York, where it was handled by the fruit and produce commission houses. There, among turnips and apples and squash, cranberries were sold to the shopkeepers of the region. From the sale the merchants took a commission and returned the balance to the grower. It had been this way since wild cranberries came to market. When the first cultivated cranberries appeared, commission merchants sought out the growers of this popular fruit. In early September as the harvest began, some merchants came to the Cape, visiting growers they knew, meeting others, dickering to handle the crop on the vines. Business might be conducted at the bog or over dinner in the grower's house. The yearly arrival of these gentlemen always provoked excitement among the cranberry-growing families, for the visitors were different and urbane and brought news of a wider world than most here would ever know.

Other commission merchants, perhaps from farther afield, let a postcard do the work. Some of these were merely prosaic bulletins in which current prices were quoted and a request made for shipment of cranberries. But others were more newsy and nearly literary, like the message from Redfield and Son of Dock Street, Philadelphia:

November 20, 1895
Dear Sir,
The arrival of cranberries continues very light, and with favorable weather, an active demand and the near approach of the holiday trade, we believe that immediate and regular shipments will pay you well to make to this market, and anything dry and solid will sell as follows: Fancy, $10.00 to $11.00 per barrel. Choice, $8.50 to $9.50. Medium grades, $7.00 to $8.00. Ordinary stock and that showing wet and weakness, selling at from $6.00 to $7.50 per barrel, as to quality and condition.

As the Thanksgiving and Christmas holidays passed, so did the traditional time of greatest cranberry consumption. Accordingly, the message from Redfield and Son changed:

December 29, 1888
The holiday trade being over and well supplied, we look for the usual dull trade for the next two or three weeks and would advise you to hold your shipments until we advise you, which will be as soon as there is any demand for the same.

So it went. Most growers preferred to move their berries from September through December, when demand was the greatest. The Early Blacks handled the Thanksgiving trade, while the varieties picked in October—Howeses, McFarlins, and Centervilles, to name a few—reached the market in time for Christmas. No wholesaler could guarantee a grower a set price for his berries; it depended on the quality of the fruit, the supply and demand, and the state of the economy. But prices of between eight and ten dollars or more were regularly received, and a grower tended to sell through a favorite commission house that dealt squarely with him and consistently got a good price for his berries. Selling the crop always elicited a competitive spirit in cranberrymen. If a fellow's house had gotten him several more dollars per barrel than his neighbors had received, they and the rest of the town soon knew about it. There were always some who held their berries until late in the season, perhaps February or March, hoping for an even better price when supply was low. Sometimes they got it, but the proposition was risky. When cranberries sat for days in a storehouse, they shrank, lost weight, sometimes froze, and then rotted.

WILLIAM CROWELL,
Wholesale dealer in
Cape Cod Cranberries,
AND GENERAL COMMISSION MERCHANT,
39 Water Street,
Formerly firm of
BAKER & CROWELL.
NEW YORK.

From

Bbls.

William Crowell.
Ocean Spray Cranberries, Inc.

The way in which cranberries reached the market changed through the years. In the mid-nineteenth century, the crop was sailed south to New York, Philadelphia, and Baltimore and north to Boston in the holds of the Cape's ubiquitous coasting vessels. But in 1848 a railroad branch linked to Boston had reached Sandwich. By 1854 it was in Hyannis, and eleven years later, on the upper Cape at Orleans. Gradually the branch lines multiplied into more towns, the railroad became the standard, most efficient means to move the fruit, and it remained so into the 1950s. In odd lots with other freight or filling entire cars iced for the trip, cranberries traveled to the commission houses of Hall and Cole or York and Whitney at Quincy Market. Going to the mid-Atlantic states, the railroad shared the load with the steam packets of the Fall River Line and other companies whose vessels carried passengers and goods down the coast. At the port and rail yards of New York, the commission merchants Geo. L. Fisher, French & Co., A. M. Banks & Son, and William Crowell were there to meet the Cape Cod crop.

If there was one man whose life mirrored the rise of cranberry growing on the Cape in the nineteenth century, it was William Crowell of Dennis. Crowell was born in 1814 and lived until 1909. In that long span of years, he worked in nearly every aspect of the cranberry business, doing enough to

William E. Crowell, Dennis, Massachusetts.
Lindy Gifford photograph

He used to handle a large part of the cranberry crop from the lower Cape here. He also handled New Jersey berries. He used to have a standing offer that he would give vines free to anyone who would set them out in New Jersey and turn the berries back to him and so forth.

—William E. Crowell

be profiled as one of the great cranberrymen in the 1890 *History of Barnstable County*. But the best and most interesting life of Mr. Crowell did not come from the county history, but from his grandson, William E. Crowell, who lived in his grandfather's house in Dennis.

The younger William E. Crowell divided his professional life between lawyering and managing bogs as his father and grandfather had done. He was the third generation of the Crowells to have some involvement with the cranberry—not an uncommon thing in these parts—and kept up an acre or so which he harvested himself into his eighties. Although he was very young when his grandfather died, the family history was kept current, and William E. Crowell, an intelligent and plain-spoken New Englander, made his grandfather's times sound like the day before yesterday.

Like many who succumbed to cranberry fever, William Crowell's first work was on the water.

He had a boat that was built down in East Dennis. They used to build those schooners out of the Shiverick Shipyard. He had the Vestal, V-E-S-T-A-L. I suppose she was about sixty feet long, probably. They were small fishing schooners. They carried a crew of nine, and it was mostly hand fishing. They used to go off the Georges Bank, go out around Provincetown, Georges Bank again. In 1841 they had a terrible gale here, what they call the October Gale. There were twenty-three people drowned right in this area, this village here. He lost his brothers. There was a boat, the Bride, B-R-I-D-E, and two of his brothers were on that. They were both drowned. I think that discouraged him. He got back to Provincetown alive, but the others, lots of them, didn't.... After that he went into the fish business down here at what they call Corporation Wharf. He had a fish business down here for several years. Then after that he strayed off to New York. He went with a man named Baker from South Dennis; they got a partnership going. That was in the 1850s and '60s, and the shipping went down after a while. Ship chandlers, that's what they called them. They said he had a big brass cannon out in front of the place that was there...for advertising. That went down, and he went out of that business and he went into the fruit brokerage business after that.

So William Crowell was drawn to the rising star that the cranberry industry was becoming for Cape people, and, drawing on his contacts with them, he opened a fruit brokerage house known by his name. And as his grandson recalled:

He used to handle a large part of the cranberry crop from the lower Cape here. He also handled New Jersey berries. He used to have a standing offer that he would give vines free to anyone who would set them out in New Jersey and turn the berries back to him and so forth. He was much interested in the cranberry business.

It was in autumn, at harvest time, when William Crowell would find his way back to Cape Cod along with other fruit brokers.

He had a horse and buggy, and he'd go around and visit the various people and often inspect the berries that they'd already harvested.... And of course, the old-timers, they screened out their own berries, and sorted them and put them in barrels. And they had what they called a truck cart and a horse. And they'd put maybe a dozen barrels at a time on the truck cart. And the refrigerator car would be lying over at Yarmouth or Brewster or some other place. And they would go to the station and they would put them into the car. And usually more than one person would be loading. He might have bought half a dozen small crops in this area, or something. And probably within a week they'd have the car all loaded up. Two hundred barrels, I think the usual freight load was.

Like most brokers, William Crowell took the cranberries mostly on consignment.

They were trusting souls in those days, and he had a reputation. And some of them I guess he bought outright, but a great many were consigned to him, and he took a commission on his sale and remitted the balance to them. There was one time, I don't know just when it was, when there was a financial panic and paper money wasn't much good. And he brought up a thousand silver dollars in a bag from New York. And he wouldn't let the porters carry the bag; he carried it himself. He said it was pretty heavy with a thousand silver dollars in it.

Brokering the crop was not William Crowell's only foray into the cranberry business. Like so many, he was concerned with making the cranberry harvest more efficient, and in September of 1874 he filed a patent for a cranberry-picker of his own design. It was an ambitious invention, in which a man pushed a large rectangular box on wheels through the vines as an automatic rake captured the berries—truly a picking machine. Though it was displayed at a meeting of the Cape Cod Cranberry Growers Association in 1888, Crowell's cranberry-picker was seemingly soon forgotten. Four years later, he patented a more modest cranberry-picker, akin to the snap machine, with the help of Luther and Zebina Hall, both cranberrymen and Dennis natives. And then Mr. Crowell brought forth a third creation. His ventilated cranberry barrel was designed so that air could circulate around the fruit and retard spoilage during shipping and as the berries were sold. Though none of William Crowell's inventions ever became popular among cranberry growers, they remain a testament to his curiosity, intelligence, and resourcefulness.

Certainly William Crowell made little if any money from his inventions, but there were ways to profit from cranberries besides wholesaling them. Knowing the full value of the berry in the marketplace, he began building bogs in the 1880s through the sale of shares.

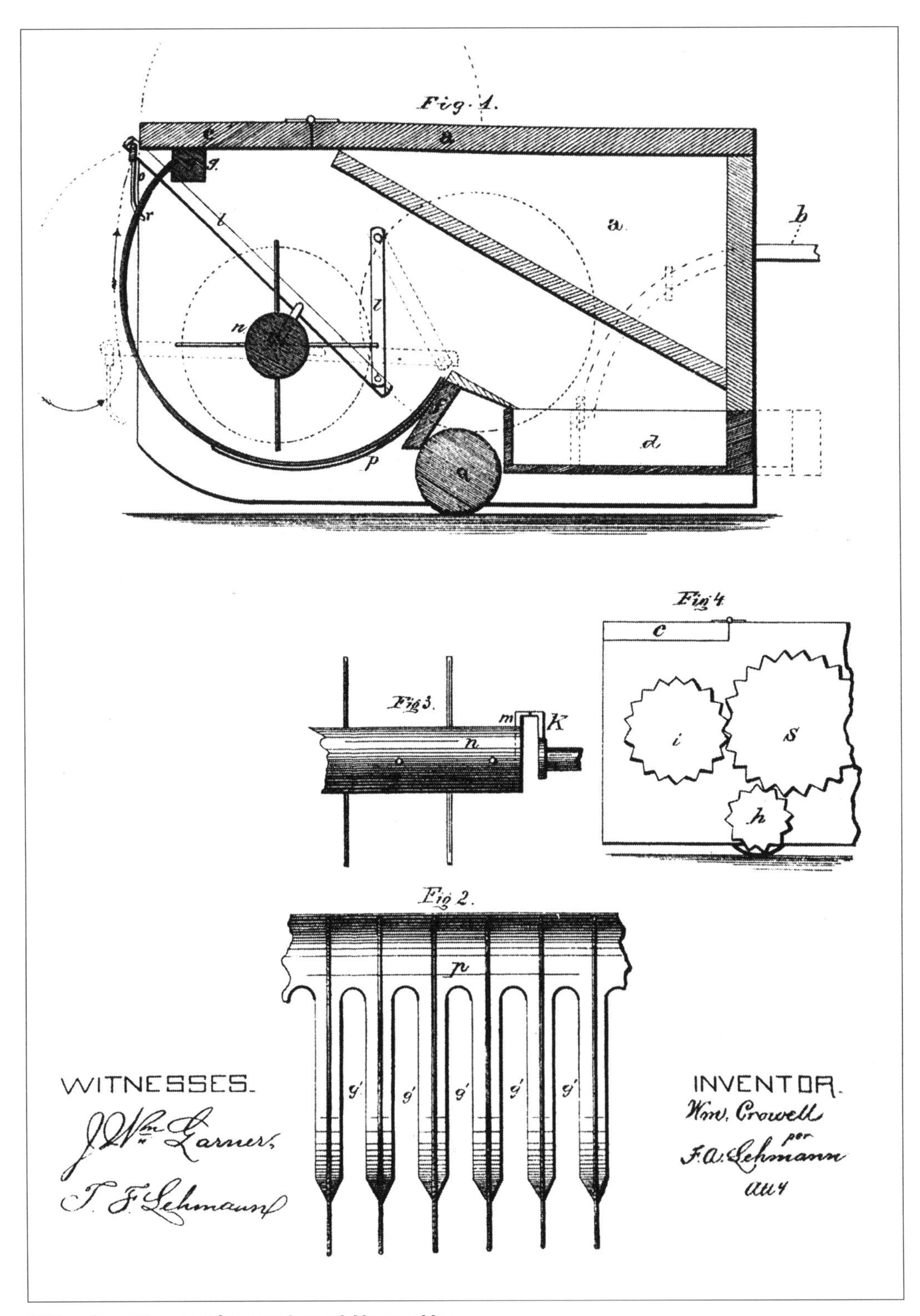

William Crowell's patent for a cranberry picking machine.
United States Patent and Trademark Office

He built a bog in Harwich known as the Great Swamp, and sold it out completely. And then he built three bogs in Wareham, which were those I was connected with afterward—what they call the Locke Bog and the Harlow Brook Bog and the Old Orchard. And they sold them out as shares, in sixteenths, thirty-seconds, and sixty-fourths, the same as the old ships used to be owned.

Crowell's Wareham bogs, in Plymouth County, were managed for fifty years after his death by his son. And when the son died, the builder's grandson, William E. Crowell, ran the bogs for another decade until the cranberry business hit rock bottom in the late 1940s. In the last decades of the nineteenth century a number of progressive cranberry growers—William Crowell among them—built bogs off-Cape in Plymouth County. Why had they chosen to locate them at a distance from the industry's heart, the place where cranberry cultivation had developed and was understood? Mr. Crowell's grandson knew the reasons well.

In the first place, the area that was suitable for cranberry bog was pretty much built up on the Cape in those days. And secondly, they had those big, what they call maple or laurel swamps up in Wareham and Carver and that area. And they were very well suited to cranberry cultivation without much work.... And the weed problem is less in that area than it is on Cape Cod here. Poison ivy, laurel, and some of those other things that bother on the Cape here are not as prevalent in Wareham and Carver and that area. They're more weed-free.... And of course they had water privileges up there that we don't have. It was possible to establish reservoirs, and to flood bogs and put [on] frost protection and insect protection and so forth. And many of these bogs around here are what they call dry bogs. Either they don't flood at all or they only flooded in the winter season, with the winter rain. And that was another reason they moved up there.

The final glacier of the Pleistocene made Cape Cod a landscape of moraine—low hills formed of boulders, sand, and gravel—with a lesser amount of level, sandy outwash plain. Fresh water occurs as hundreds of small, deep ponds with little surrounding wetlands. In general the land is poorly drained. The number of rivers and small streams on the Cape that provide good cranberry habitat and a constant water supply can almost be counted on one hand.

Northwest of the Cape, in Plymouth County and parts of adjacent Bristol County, conditions are different. The land consists of broad and flat outwash plain, made of gravel and sand washed from the base of the receding glacier, with a smaller proportion of morainal hills. And here, glacial action produced enormous wooded swamps cut by streams and rivers that wandered to the coast. The result is an optimum cranberry-growing landscape with miles of peaty wetland easily converted to bog, enormous sand deposits, and natural waterways for flooding or draining bogs.

Cranberry growers were not the first to consider the swamps of southeastern Massachusetts an economic resource. In the eighteenth century, bog iron was found in these same wetlands and became the genesis of a great industry in Plymouth County. The region's extensive pine woods provided the charcoal needed to smelt local ore into iron in the blast furnaces found at Middleboro, Rochester, and Carver. The end results of iron manufacture were tools cast at nearby foundries and nails cut at small shops established for that purpose. This use for the great wetland tracts petered out in the early nineteenth century when ores with higher iron content were discovered outside New England. Though ore was no longer extracted from the swamps, the well-established iron industry continued and grew. New ores were shipped to coastal Wareham and transported to iron mills which now primarily produced nails, shipping them by the millions throughout the country.

Into the wetlands and swamps, now dormant, came the cranberry industry, a circumstance of remarkable economic rebirth. It was an instance of rare luck for these Massachusetts towns, for seldom are the resources for two distinctly different industries found in the same natural environment. Of course, the fruit itself was no stranger to Plymouth County, having long thrived in these same lowlands and been harvested at places like New Meadows, the 500-acre freshwater wetland in the southern part of Carver. And in a small way, cranberry cultivation had been coming along, too. Every town in the county could claim some cranberry experimenters and cultivated bog. But in the late 1870s, these efforts were dwarfed by the scale of cranberry cultivation that arrived in Plymouth County.

The grower who led the industry off Cape Cod, building the most acreage in those early years, was Abel Dennison Makepeace. Imbued with considerable entrepreneurial spirit, Makepeace brought it to bear on the cranberry. He had started humbly enough as a harness maker, moving from Middleboro to Hyannis, Cape Cod, in 1854. There, on a small farm, he grew potatoes, rutabagas, and strawberries, eventually building an acre or so of bog in a hollow on the property. Apparently, Makepeace's earliest cranberry investments lost money, a circumstance that rarely occurred in his later ventures. From a shred of backyard bog, Makepeace, thrifty and shrewd, went on to buy land and water rights on Cape Cod, build some bogs, and buy others. The capital he used was not entirely his own, for a Boston financier named George Baker became his backer, believing that Makepeace's bogs would produce far more capital than it had taken to construct or acquire them. The properties called Folger, Newtown, and Santuit, to name a few, disappointed no one. The partnership of Makepeace and Baker began a trend of growers and lenders joining forces to amass many bog properties.

Makepeace began work in Plymouth County in 1877, laying out in Carver, appropriately enough, the 60-acre Carver Bog. In the same town in 1882, Makepeace put in a 100-acre bog known as Wankinko, and three years later built the 77-acre Frog Foot Bog, less than a mile to the north. These properties were enormous compared to what had come before them. On Cape Cod, Makepeace never owned a bog larger than 40 acres. In order to finance the new bogs, Makepeace sold shares in them to individuals, forming ownership associations for each property; thus the Frog Foot Bog was owned by the Frog Foot Company.

By 1890 Makepeace's success in the cranberry industry had been so great that he was referred to quite simply as "The Cranberry King." The leading grower of his time, Makepeace was also president of the Hyannis National Bank, owner of a general store in West Barnstable, treasurer of the Barnstable Brick Company, and a backer of the Barnstable County Electric Railroad. But despite all this, Makepeace never lost a strong Yankee frugality. His son John, eventually president of the A. D. Makepeace Company, once told a friend that when he was young and his trousers had worn through at the knee, his mother would cut the legs off, turn them around, and sew them back on again.

All the while, Makepeace was called upon by others to oversee construction of their bogs and lend his considerable experience to their enterprises. In 1885, partly due to one man's efforts, Plymouth County supported 1,347 acres of cranberry bog, already a little more than half of Barnstable County's 2,400 acres. The town of Carver became the center of Plymouth County cranberry growing, blessed as it was with seemingly more swamp than

A. D. Makepeace Co. screening crew, West Barnstable. In the front row, Abel Dennison Makepeace is second from right. His son, John C. Makepeace, is at the extreme left.
A. D. Makepeace Co.

upland. Where A. D. Makepeace had developed so many properties, others did as well. The sizable Bowers and Russell Bog was built at Carver's East Head in 1878. Later in the century, Federal Furnace Cranberry Company was established with large bog holdings on the site of a former iron furnace. Within this same part of town, South Carver, the substantial Shaw and Slocum–Gibbs bogs were established in the early years of the 1900s. In 1890 the land in Carver devoted to cranberry growing amounted to 750 acres. Twenty-two years later, in 1912, there were 2,461 acres of cranberry bog in town. Carver and Plymouth County had displaced Cape Cod as the heart of the cranberry industry.

The word plantation would soon be linked with some of the county's largest cranberry companies. While they lacked the economic self-sufficiency of Southern ante-bellum plantations and the institution called slavery, there were general similarities. The big companies were often self-contained and their social structure evident on the landscape. The Federal Furnace Cranberry Company, which raised more than a hundred acres of cranberries in an iso-

Screenhouse, Federal Furnace Cranberry Company, Carver, Massachusetts.
Lindy Gifford photograph

Worker housing, Federal Furnace Cranberry Company, Carver, Massachusetts.
Lindy Gifford photograph

lated location, had several homes for foremen and supervisors, an enormous screenhouse with various outbuildings, and seventeen tiny worker shanties. In modest New England, these farms with their diverse buildings and populations and huge land holdings were the closest equivalent of the Southern plantation. As we shall see, the cranberry plantations—consciously or unconsciously—fostered a certain dependence among their migrant, often black, workers.

Beyond its own burgeoning industry, Carver became an exporter of cranberry culture through the efforts of two native sons, the McFarlin brothers. In 1870 Thomas Huit McFarlin took from the great natural bog called New Meadows a strain of vine and berry he thought promising. The McFarlin berry thrived in local cultivated bogs and soon traveled farther than its discoverer. Thomas's brother, Charles McFarlin, was the wanderer in the family. He had gone to California during the gold rush, returned to Carver in the 1870s, and learned about the culture of red gold—cranberries. Soon enough he was off again, this time to the Pacific Northwest and the vicinity of Coos Bay, Oregon. In 1885 Charles McFarlin sent home for McFarlin cranberry cuttings. With them, he helped to start the modest industry of small, family-owned cranberry bogs that still continues on the Oregon and Washington coasts.

That Carver and the surrounding Plymouth County towns would remain central in the culture of cranberries was confirmed in 1909 when the Commonwealth of Massachusetts chose to locate a cranberry experiment station in East Wareham. Beginning as two buildings set near a pond in a remote part of town, the station soon included its own cranberry bog. To direct the station and cure the cranberry's ills, the Commonwealth sent Dr. Henry J. Franklin, a specialist on insects who happened to be the world's expert on the bumblebee. Franklin's job was to apply science to the growing of cranberries, whether this meant identifying and combating predatory insects or determining the effects of weather and water on bogs. So committed was Franklin to the task that he actually lived at the station during the first seven years of his appointment. As future chapters will reveal, Dr. Franklin assisted growers immeasurably during his forty-seven years as director. So remarkable was his work that when he died in 1958, both the *New York Times* and the *New Yorker* would note his passing. This was no small honor for a scientist who dedicated his life to the study of fruit cultivation.

While the fortunes of cranberry growing in southeastern Massachusetts rose, they began a long, slow decline on the Cape. Cranberry acreage peaked there in 1905 at 4,700 acres and would fall off gradually for years to come. The diminished productivity of bogs built on unsuitable land, a disease called false blossom, which damaged nearly every Cape Cod bog, and lack of interest in cranberry growing by the coming generation all contributed to the cooling of cranberry fever. A traditional story circulated among Cape growers was of the man who inherited a bog from his relations. About mid-August he would visit the bog to size up the prospects, his first visit since last year's harvest. If he saw a good crop, he'd scythe down the grass in preparation for picking. If the berries were sparse, he left the bog alone for another year. Yet if Plymouth County had become the heart of cranberry growing, Cape Cod remained its soul. After all, the fruit were still known universally as Cape Cod cranberries.

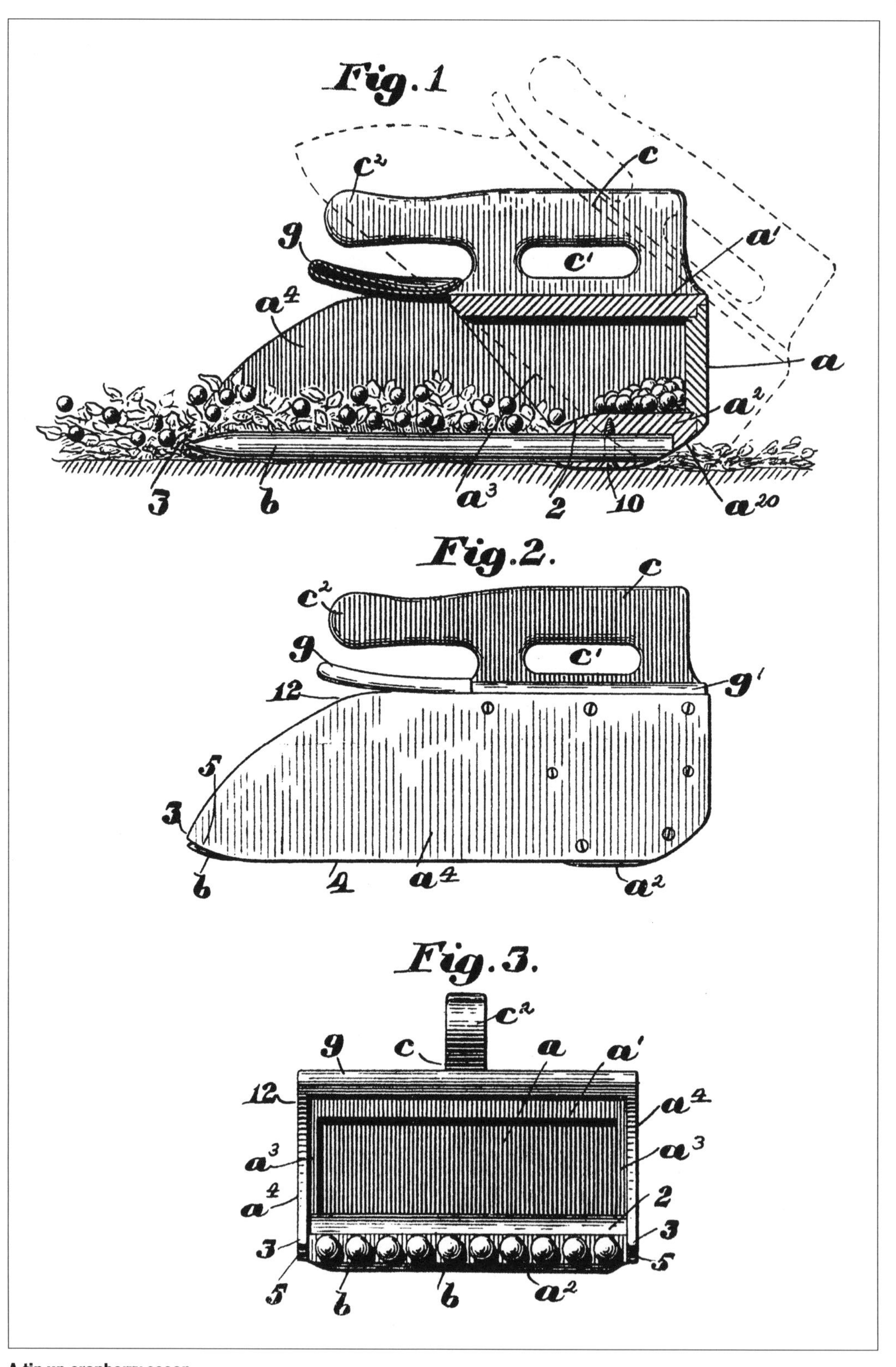

A tip-up cranberry scoop.
United States Patent and Trademark Office

Chapter 4

Wooden Age Tools

Since cranberry growers had undertaken a new kind of agriculture, it followed that they needed new tools for cultivation. Farmers growing grains and vegetables inherited from earlier generations basic growing methods and tools; implements like sickles and hoes had been around in various forms for thousands of years. The cranberry grower had no such luck. The centers of cranberry cultivation were remote from those of industry where new technology was developed, so it was unlikely that any agricultural manufacturer could know enough about cranberry growing to develop tools that filled the farmer's needs. And too, the size of the industry was minuscule when compared to dairy farming or even apple growing. The cranberry industry was too small to bring a large toolmaker any real profit.

Left to their own devices, cranberry growers devised a number of different implements by the 1880s, mostly for building bog. There was the turf axe, a primitive-looking item with a thin steel blade and long hickory handle. It was used to cut turf when building bog and to trim the growth along the ditches once the bog was established. The turf axe's companion was the hauling rake—two steel talons attached to a handle—which aided in manipulating the turf off the bog and onto the dike, of which it became a part.

Some very simple tools also were invented for preparing and planting bog. The spreader and marker were near twins. The first was used to spread sand over the peat and consisted of a long handle with a narrow board on the end. At a glance, the marker seemed to be an enormous rake. The handle was nine feet long and at one end was a two-by-four-foot joist in which eight-inch white oak teeth were set eighteen inches apart. Drawn over the sanded bog north to south, then east to west, the marker made a checkerboard pattern, and vines were planted at the intersections of the lines. The vine cuttings went in with another tool of white oak, the setting stick. It was simply a stick eight inches long with a rounded handle, used to press cuttings down into the sand.

Each of these tools was invented within a period of about forty years, some, no doubt, earlier than others. All resulted from a cranberry grower asking himself how he could make a certain task easier and then heading for his workshed to try. If solving the problem required tools he did not own or skills he did not have, the grower went to a neighboring carpenter or blacksmith for help. In

Preparing a bog and setting vines. Massachusetts Cranberry Experiment Station

towns far apart, cranberry growers devised the same tools for the same job, unbeknownst to each other. But each fashioned it a little differently, so that many variations on one theme were made. Such was the case with the setting stick. Besides the form described earlier, some setting sticks were merely a long wooden peg narrow at the bottom, others, pistol-shaped with a metal shaft. Growers also used narrow, sawed-off shovel handles to plant vines. Not until the early twentieth century, when small shops started to produce equipment, would tools for growing and harvesting really become standardized. Before then, they were mostly items of folk culture, produced not for sale, but for home use, and made by the user. What was true in the nineteenth century is still true today: the tools of the cranberry trade are made in the growing region. And then, as now, the progress of cranberry culture depends on the ingenuity of growers, bog workers, and local craftsmen.

The most important tool of all in the cranberry trade was the last to come into common use—the harvester. Hand picking had always been the way, both on wild and cultivated bogs. But toward the end of the nineteenth century, a problem became apparent. Because of good returns, cranberry acreage was growing quickly on Cape Cod and up in Plymouth County, more quickly than the supply of harvest laborers. There were two ways to approach the problem: find another labor supply or develop and use a cranberry picker. Belatedly, cranberry growers did both.

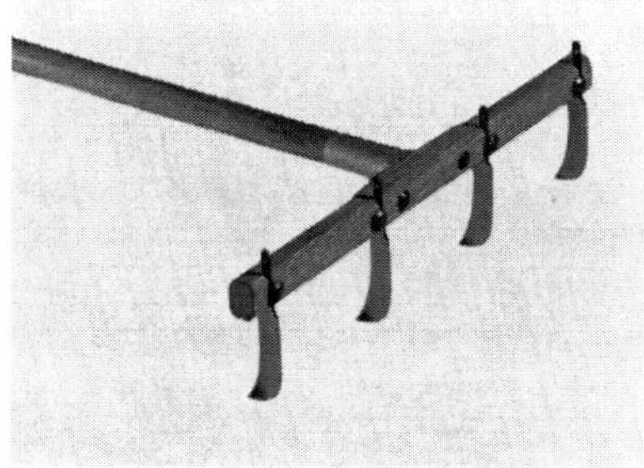

BAILEY'S
Knife Rake

● For pruning vines. Four removable curved blades; 6-foot handle.

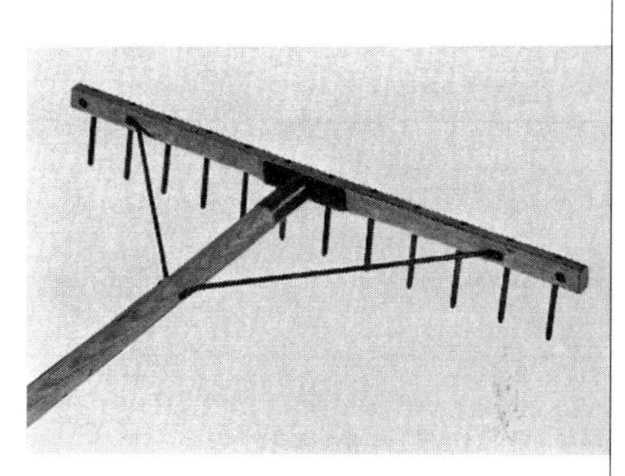

BAILEY'S
Metal Tooth Vine Rake

● For removing loose vines after harvesting. 24″ head; metal teeth. 6′ handle.

BAILEY'S
Combination Rake

● Vine and pruning rake in one.

BAILEY'S
Turfing Axe

● For cutting turf, trimming ditches, and similar uses. Blade made of best steel. Best quality hickory handle 36″ long. Weight about four pounds.

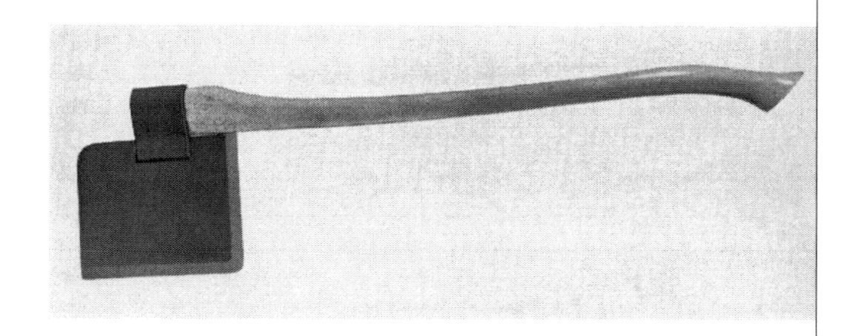

H. R. BAILEY COMPANY *Equipment for the Cranberry Grower*

Manufacturers of SEPARATORS · BLOWERS · ELEVATORS · BELT SCREENS · CONVEYORS · BOX SHAKERS · BOX PRESSES SCOOPS · SNAP MACHINES · HARVESTING BARROWS · VINE SETTERS · WEEDERS · SAND BARROWS · SAND SCREENS TURF AXES · TURF PULLERS · KNIFE RAKES · VINE RAKES · PUMPS · DUSTERS · GAS LOCOMOTIVES

SOUTH CARVER **Telephone 28-2** **MASSACHUSETTS**

Courtesy Eunice Bailey

BAILEY'S
Steel Box Press

● For pressing covers on one-quarter and one-eighth barrel boxes. Ball-bearing head revolving for easy nailing. Self adjusting for slight variation in boxes.

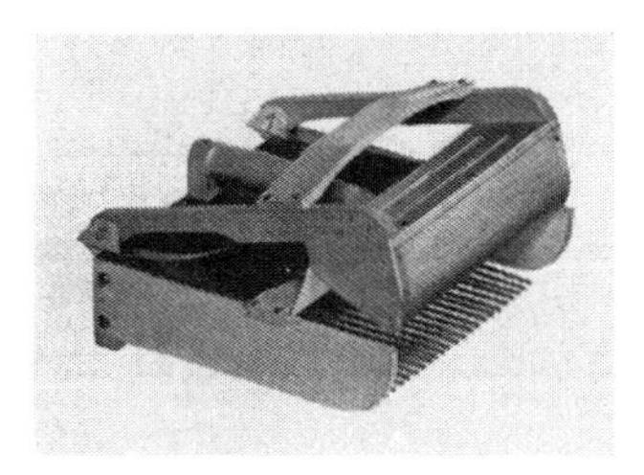

Snap Machine

● Made in several sizes — 24, 26, 28 and 30 steel teeth. For picking berries on young, short or tangled vines.

BAILEY'S Cranberry Scoop

● Curved wood-tooth scoops. Metal back. Wire screen top. Raised handles. Standard size, 21-tooth. Other sizes to order.

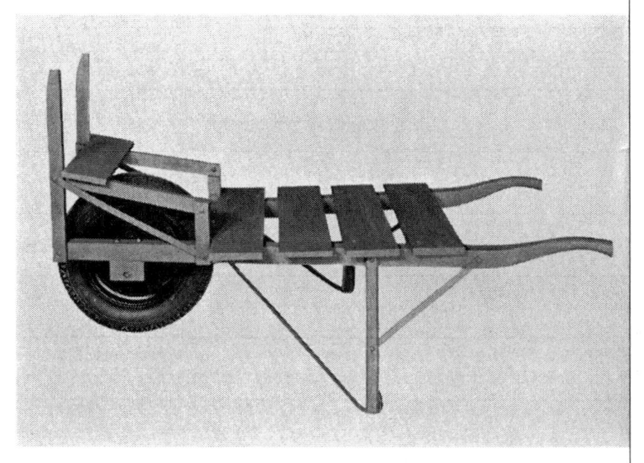

BAILEY'S Harvesting Barrow

● For carrying boxes. Has 20″ x 4″ double-tube pneumatic tire and roller-bearing wheel. Drop handles.

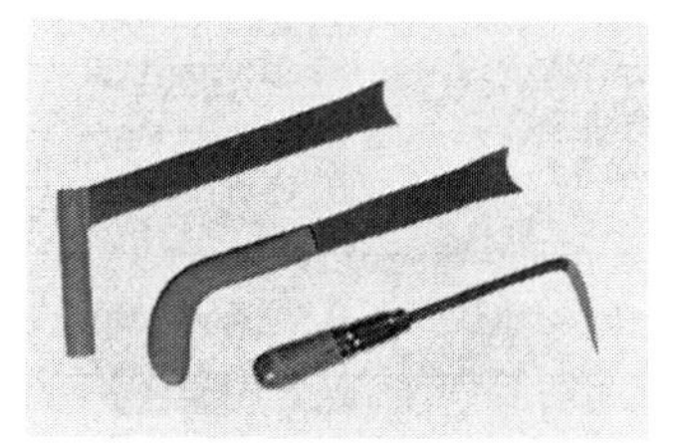

BAILEY'S
Vine Setters and Weeding Hook

- Vine Setter all steel.
- Vine Setter wood handle, steel blade.
- Weeding Hook.

H. R. BAILEY COMPANY *Equipment for the Cranberry Grower*

Manufacturers of SEPARATORS · BLOWERS · ELEVATORS · BELT SCREENS · CONVEYORS · BOX SHAKERS · BOX PRESSES SCOOPS · SNAP MACHINES · HARVESTING BARROWS · VINE SETTERS · WEEDERS · SAND BARROWS · SAND SCREENS TURF AXES · TURF PULLERS · KNIFE RAKES · VINE RAKES · PUMPS · DUSTERS · GAS LOCOMOTIVES

SOUTH CARVER **Telephone 28-2** **MASSACHUSETTS**

Courtesy Eunice Bailey

In 1883 a paper entitled "Cape Cod Cranberry Methods" by Mr. O. M. Holmes of Boston was read before the American Cranberry Growers Association, an organization of New Jersey growers. Under the heading "Machine Needed," Holmes wrote:

> A machine that would collect ninety percent of the berries would be a perfect "bonanza" to the inventor, and the grower could well afford to pay a good price for it, or a royalty that would make it an object for some inventive Yankee or Jerseyman to produce. It seems very strange to me that, with our natural inventive genius, there has not been a cranberry picker invented before this. Why cannot we put our heads and pockets together, and work this machine to an actual reality? Why not offer a reward to develop our natural inventive genius, in the shape of a cranberry picker? We have reapers, potato diggers, and hundreds of mechanical inventions that are far more complicated. Such a machine would not seem to be impossible.

It was not the failure of America's "natural inventive genius" that allowed cranberry farmers to arrive at the year 1882 still picking their crop by hand. Manufacturers and inventors concentrated efforts on products that served the many, not the few. If things had been otherwise, we might also have had a cranberry picker in 1846, when the sewing machine was invented.

First, for every cranberry-growing tool, demand was small. As it was, almost all the men who held patents on cranberry pickers in the nineteenth century were from Cape Cod or Plymouth County and were, it is likely, growers. Since at least 1865, this group—with old local names like Hall, Thatcher, and Crowell—had turned out a diverse array of picking inventions. Some pickers were probably designed before 1865, but since earlier patent records were lost in a Washington fire, we will never know their identity. So cranberry pickers *existed* in 1882 when Holmes made his plea, but were poorly known, since few ever made it to commercial production. Surely other picking contraptions were kicking around, too, made by inveterate tinkerers to try out on their own bogs and those of friends. These men had neither the aspirations for their inventions nor the level of sophistication that a patent implied. While there was plenty of inventive genius among individual growers, cranberrymen as a group were wary of change and offered little encouragement to those with new ideas.

The search for a cranberry picker took three paths. The evolution of the picker known as the snap machine began in the 1870s and was an attempt to imitate the motion of hand picking. No one earning money picking cranberries ever gathered them one by one, as strawberries are harvested. The standard method consisted of trapping some vines near their base by the edges of each hand or the fingers pressed together. Then the hands or fingers were drawn up and back over the vines, leaving in them the berries. The snap machine did much the same. It was a small wooden box with metal tines on the bottom and a movable upper jaw. When the picker was thrust into the vines, the jaw was snapped downward with the thumb, trapping fruit between the tines. It was

---

It seems very strange to me that, with our natural inventive genius, there has not been a cranberry picker invented before this. Why cannot we put our heads and pockets together, and work this machine to an actual reality?

—O. M. Holmes,
"Cape Cod Cranberry Methods"

Hand picking.
Middleborough
Public Library

then pulled out backward with the bounty. Of course, variations were developed. In one, the upper jaw was lined with wire bristles resembling a mustache, as opposed to the traditional wire mesh or wooden jaw. In another, the upper jaw had been entirely removed and the picker used a small brush to sweep berries into a receptacle which was held like a dustpan.

The rather modest snap machine, with its single moving part, was perfected in 1883 by Daniel Lumbert from the Cape village of Marston's Mills, but it took a while to catch on, not being used with any regularity until the 1890s. What made growers cautious was concern for their vines. Most crops are

The snap machine.
United States Patent and Trademark Office

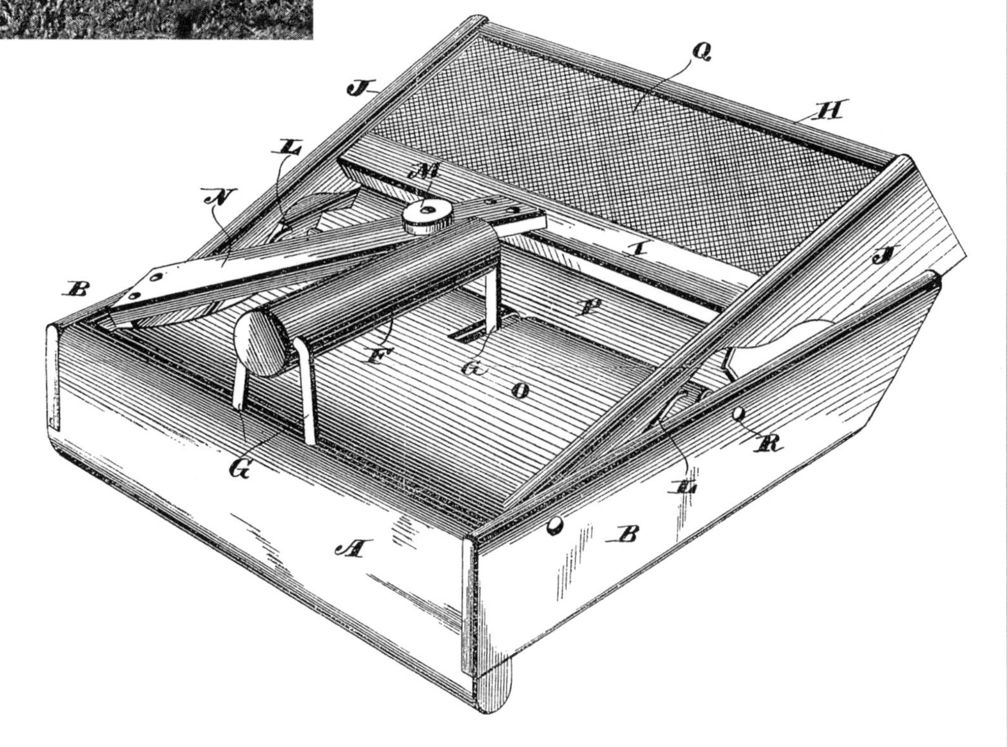

annuals and the plant is destroyed in the harvest or shortly thereafter. But the vines that held the cranberry crop would hold many crops to come if treated well. At first, cranberry growers did not believe that "snaps," as they were called, could work without pulling up vines. But slowly they came to see that the snap machine was nearly as careful as the human hand and a good deal faster.

A second tack, taken by the most ambitious of men, was to produce an actual cranberry picking machine. All had moving parts, were drawn or pushed along on wheels, and were part of the trend toward agricultural machinery begun in the 1830s with the mowing and reaping machine and thresher. An element common to almost all the inventions were tines, the agent of harvest, which either remained stationary or revolved through the vines. Behind the tines was a large receptacle for the berries, beneath which was a single axle and wheels.

These harvesters first appeared as patents in the late 1860s and are found regularly thereafter. Few ever got beyond the experimental stage and none saw regular use on a cranberry bog. The problem seems to have been the reliance on manpower to push the harvesters along. Horses, the source of power for other agricultural machinery, would have damaged bogs and vines with their hooves, and were therefore out of the question.

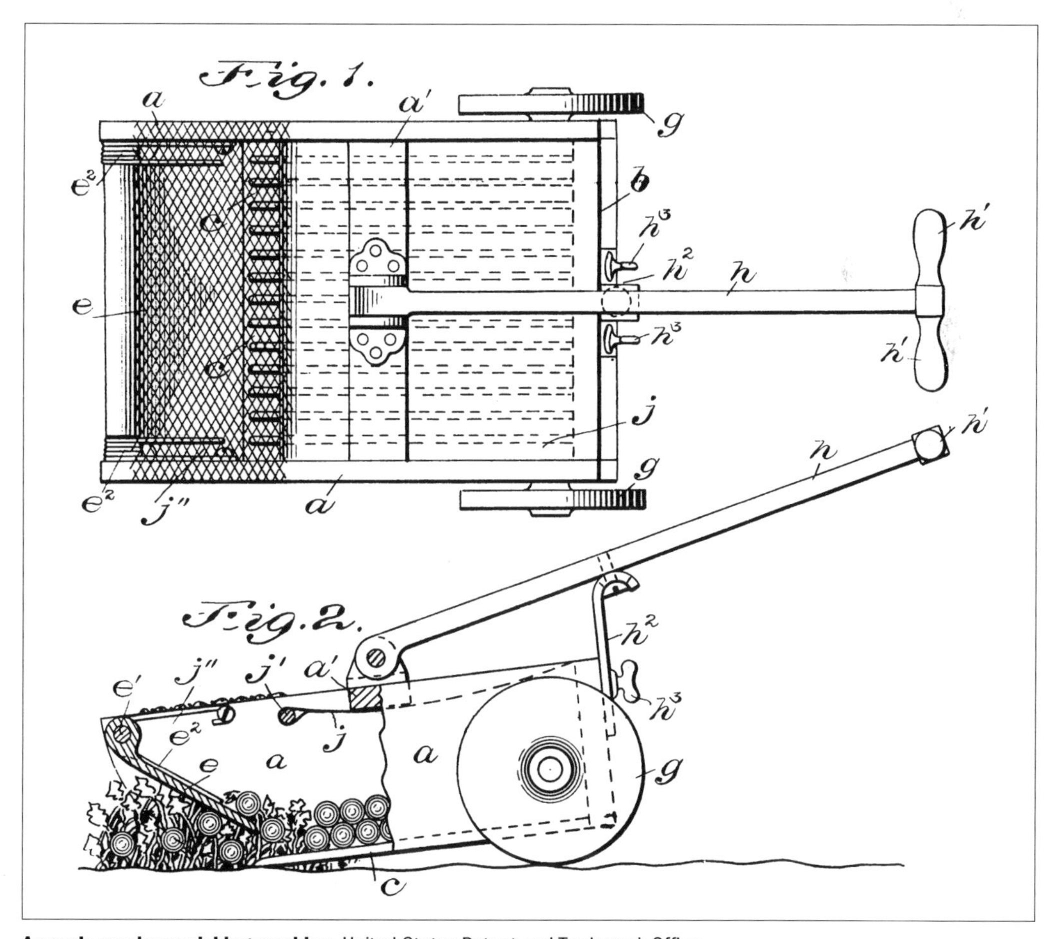

An early cranberry picking machine. United States Patent and Trademark Office

But no man could provide the consistent flow of energy necessary to move a large wheeled machine through the dense mat of a cranberry bog without tearing up the vines. An added problem was the fact that most cranberry bogs were anything but level. Early machines could not be adjusted for changes in elevation, so they often chewed up the bog. It could have been that a gifted inventor might have devised a light and workable harvester powered by man, but in the early years of cranberry growing, this never occurred. What remains striking, however, is that in design, several of the early harvesters prefigured the successful gas-powered pickers perfected in the 1950s.

In the end, cranberry growers chose the simplest picker to use as their harvesting tool and the one they had first experimented with. It was, of course, the cranberry scoop. Almost primitive in form, scoops look as if they might easily have been invented in the early eighteenth century. Once again, the first known designs of cranberry scoops date from 1865 (after the U.S. Patent Office conflagration). The scoop looked like a comb to which sides, a top, and a handle had been added. While the snap machine was pulled backward out of the vines, the scoop was pushed into the vines, then up and out. The harvester worked on his or her knees, or bent at the waist, to scoop.

Cranberry growers were even more suspicious of the scoop than they had been of the snap. The motion of combing through the vines, they felt sure, would rip out vines and prove rougher on the berries than hand picking. And it was found that while 10 percent of the cranberries on the vine was lost in hand picking, a full 20 percent was lost in scooping. For some growers this extra 10 percent seemed too great a loss. But by the end of the nineteenth century, most cranberrymen were coming to favor the scoop. In 1895 the Cape Cod Cranberry Growers Association met at Middleboro's Grange Hall to consider the "merits and demerits of the so-called Scoopers." They grudgingly admitted that when used "under proper restrictions, they were of great advantage and profit." The great advantage of the scoop was speed. Pickers could scoop a bog in less time than it took a far larger number of hand pickers to harvest it. For the grower, that meant paying out less money in wages and a better chance of getting his crop in before the frosts came. At the start of the new century, the scoop was in regular use on Massachusetts bogs. The snap, which held less than a quart of berries to the scoop's quart and a half or two quarts, was now relegated to use on young bogs. In places where the plants were not yet well rooted, this more delicate tool gathered the crop with less loss of vine.

> ***The scoops were so different, I don't know in a group of pickers if there would be two scoops alike. There were many of them, they were homemade.***
>
> —Malcolm Ryder

A funny thing happened to the cranberry scoop as time passed: it grew larger and larger. The earliest ones were little bigger than a snap machine, perhaps eight inches in width, and were held with one hand. Cotuit grower Malcolm Ryder remembered these scoops well from the first years of the twentieth century, when, as a youngster, he would work and play on his father's Cape bogs. "But the scoops were so different," he said, "I don't know in a group of pickers if there would be two scoops alike. There were many of them, they were

Harvesting with a rocker-bottom scoop.
Middleborough Public Library

homemade." It was so. Cash was always scarce on the remote and rural Cape, and the manufacture of cranberry scoops was still on a very small scale. A person could build a scoop as easily as buy one. Certain items were prized for the job, like the boxes in which plug tobacco was shipped to the stores. Wood from the boxes happened to make excellent scoop teeth when cut and shaved down.

The small homemade scoop soon grew to a foot in width, becoming a much heavier tool. Its handle was extended so that it could be worked with both hands. The larger a scoop, the more berries it could pick and hold, and the faster a bog could be harvested. Generally this scoop was known as the tip-up in the early twentieth century; with long straight teeth, it was necessary to tip the scoop up at the end of each motion to part the berries from the vines and the vines from the teeth.

Toward the 1920s the tip-up was replaced by the last successful innovation in cranberry picking for the next thirty years: the rocker-bottom scoop.

Ernest Howes, Wareham, Massachusetts.
Lindy Gifford photograph

The rocker-bottom had been patented as early as 1900, but, as always, it took time for it to find acceptance. The base of the scoop was rounded and the teeth curved upward; it was pushed through the vines with a distinctive rocking motion that allowed the picker to get in and out of the vines without too much difficulty. It was a great deal larger than the tip-up, from a foot and a half to two feet across with twenty to twenty-four teeth. When filled to absolute capacity, the largest of the rocker-bottoms held half a bushel of cranberries. The rocker-bottom scoop was the end of the line, as full-blown and efficient as a simple hand tool could possibly be, built to keep up with the ever-increasing cranberry acreage.

Rocker-bottom scoops were made all over Plymouth County once they caught on, and each maker's scoop had slight patented variations. In Hanson, the Eversons were scoop makers; at Marshfield, the label on the scoops said G. H. Chandler. W. B. Waters turned out rocker-bottoms closer to the heart of the growing region in the coastal village of Manomet. In the very thick of Carver's bogs, H. R. Bailey had scoops for sale, while next door in Wareham, the growers A. D. Makepeace Co. made and sold their own.

There was no rapid production line in scoop making, only a craftsman and an assistant or two, slowly making handsome wooden tools with more than a little artistry. The late Ernie Howes of Wareham was one of these woodworkers. In the winter, when most cranberry work had slowed to a crawl, he built scoops for A. D. Makepeace Co. Makepeace's scoop shop stood on Wareham's Main Street and it was home to an elderly cabinetmaker named Mr. Briggs, and to Ernie. The wood they used was hard New Hampshire rock maple, some of it cut from sugarbush, the drill holes from the taps still visible. Mr. Briggs sawed out the scoop parts: back, sides, top, handle.

And he also had a shaper to form the teeth. And it was difficult making those wooden teeth. I'll tell you why. You saw them all out and they're uniform, but before two or three weeks was out, some change come there. Because the wood was not dry. But of course, when you open up a piece of wood, I don't care how dry it is, you're going to find moisture, and it's going to move.

It was Ernie's job to assemble the pieces, to put the scoops together.

And I looked in my diary the other day and I see where I assembled a scoop in fifteen minutes. And Mr. Briggs, his name was, he said, "Ernie," he says, "you've done a good job today."

The scoops the men made cost twenty-five 1920s dollars, and it is this fact that best testifies to the quality of the tool and the skill of the makers.

In the long succession of implements used to grow and harvest cranberries, one stands above all others for sheer ingeniousness: the cranberry separator. The separator did away forever with the need to sort the good berries from the bad entirely by hand in packing the crop. In the lore of cranberry growing, the idea for the separator originated in the New Jersey barrens. There "Pegleg" John Webb stored his berries in a loft and packed them below. Owing to the difficulty his wooden limb caused in moving berries between floors, John Webb once dumped a barrel of berries down the barn stairs. The rotten berries just sat there, while the good ones—firm and hard with their air sacs intact—bounced to the bottom. It is said that Webb's observation inspired men like D. T. Staniford of New Brunswick and Joseph Buzby of Moorestown, New Jersey, to build cranberry separator machines based on the principle that a good berry bounces and a bad one does not.

How these dumb Oscillating Rolls can sort, separate, and produce the results you see in the steady stream of cleaned and sorted berries is truly awe-inspiring and makes you feel the presence of the great inventor who first conceived the idea that sound berries thrown against bouncing-boards in a certain way would determine their own course in accordance with size, shape, condition, and soundness—and the manner in which they were hurled.

—advertisement, Hayden Cranberry Separator Mfg. Company

In the 1870s, separators were at work in New Jersey and by 1892, the technology had made it to South Carver, Massachusetts, where Lothrop Hayden was building some of his own design. All the machines were essentially the same: a wooden frame more than six feet high, within which was set a series of vertical hurdles. Cranberries were poured in the top and fell onto a wooden hurdle or bouncing board. Healthy berries bounced on, then over the hurdle to a succession of others below, each stepped a little farther forward, making their way to a grader before packing. The soft berries, their pneumatic quality impaired, could not jump the hurdle and fell to a waste box beneath the separator.

The concept was splendidly simple and worked almost perfectly. But the machine was not quite worthy of the histrionic claims made in its behalf by the Hayden Cranberry Separator Mfg. Company.

Cranberry separators and conveyors, Fuller-Hammond screenhouse, Easton, Massachusetts.
Lindy Gifford photograph

The Latest Scientific Triumph Will Save You Thousands of Dollars. A brief description of the greatest money-making machine ever offered to cranberry growers. Does the work of a thousand human hands. Does it quicker and cleaner and at slight expense to operate. As the berries descend over the Oscillating Rolls they are automatically thrown against the Hayden adjustable Bounce-Boards thus separating the sound berries from all imperfect or mushy berries; grades them into 1st and second grade sizes and deposits the berries in a steady stream to the Carrier Belts and thus to their respective barrels or boxes. It is at this step in the separating process at which Hayden genius has achieved its greatest fame. How these dumb Oscillating Rolls can sort, separate, and produce the results you see in the steady stream of cleaned and sorted berries is truly awe-inspiring and makes you feel the presence of the great inventor who first conceived the idea that sound berries thrown against bouncing-boards in a certain way would determine their own course in accordance with size, shape, condition, and soundness—and the manner in which they were hurled.

Either standing by itself or hooked up to the numerous attachments built to accompany it, the cranberry separator is an invention Rube Goldberg would have been proud to claim. It is tall and ungainly looking, with a bulbous drum sticking out the front and three belts and an extra-long bicycle chain attached to keep things moving. The earliest ones were cranked by hand, but soon small motors

BAILEY'S
CRANBERRY SCREENING EQUIPMENT

ILLUSTRATED above is an assembly of Bailey's Cranberry Screening Units, in a commonly-used combination, using individual motors with each unit.

The Units shown above — from right to left — are: Bailey Blower, Elevator, Separator and Grader, Double Belt Screen, Conveyor, and Box Shaker.

One Blower and Elevator will take care of two or more Separators. Conveyor may be extended to accommodate any number of Separators.

Instead of using boxes as shown in above illustration, we make Conveyors that can be attached under the Separator when two or more mills are used.

The Fan shaft of the Blower, Separator, and Elevator shafts may be equipped with ball bearings at additional cost if desired.

This equipment may also be operated from an overhead shaft.

H. R. BAILEY COMPANY *Equipment for the Cranberry Grower*

Manufacturers of SEPARATORS · BLOWERS · ELEVATORS · BELT SCREENS · CONVEYORS · BOX SHAKERS · BOX PRESSES
SCOOPS · SNAP MACHINES · HARVESTING BARROWS · VINE SETTERS · WEEDERS · SAND BARROWS · SAND SCREENS
TURF AXES · TURF PULLERS · KNIFE RAKES · VINE RAKES · PUMPS · DUSTERS · GAS LOCOMOTIVES

SOUTH CARVER **Telephone 28-2** **MASSACHUSETTS**

Courtesy Eunice Bailey

The Smith-Hammond screenhouse, Wareham, Massachusetts.
Lindy Gifford photograph

ran them. Although it was the heart, a cranberry separator was but one part of the complete screening unit to serve the grower. There was the blower, a fan that blew the leaves and litter out of the berries, and the elevator that moved the winnowed fruit to the top of the separator. Below the separator, a moving belt conveyed the berries toward the most sensitive tools possible for a final inspection—human eyes and hands. Local women scanned the moving mass for underripe berries and located occasional rotten ones by sense of touch, running their palms over the fruit. The work caused stiff necks and shoulders, but it did allow for neighborly conversation, if not eye contact.

No one builds cranberry separators anymore. The last to do so was the H. R. Bailey Company, which closed its South Carver shop in the 1960s. Hugh Bailey had arrived in the region from Nova Scotia before 1900. Naturally inventive, he had gone from smithing and shoeing horses to selling and repairing the newly popular bicycle. When cranberries got big in Carver he was ready, making snaps, scoops, sand barrows, and a host of other equipment.

If you ordered a separator and its accouterments from Hugh Bailey, the first thing he did was come measure your screenhouse. The screenhouse was the cranberry grower's evolving modification of the barn in which his crop was screened and

Interior, Federal Furnace screenhouse, Carver, Massachusetts.
Lindy Gifford photograph

The Bailey shop, Carver, Massachusetts.
Lindy Gifford photograph

packed. It was small and shed-like or large and multistoried, depending on the size of the bogs it served. Many shared a distinctive shape, an elongated roofline reminiscent of the saltbox houses of early New England. Almost all of these buildings were constructed with a long bank of windows at the south or west, providing screeners with the abundant natural light so useful in sorting cranberries. The screenhouse would become the cranberry industry's contribution to the region's architectural character. Few were of precisely the same dimensions, and the machinery had to fit the layout, so all screening equipment was custom built. Next Hugh Bailey might head to the great lumberyards of Boston. The stock he worked with was not cut in the pine woods of Carver. Into a separator went finer woods—ash and birch—that stood up well and caused the berries to bounce just so. Finally, Hugh Bailey, his son Donald, and perhaps another helper went ahead and built the cranberry separator—from scratch.

When we visited, little had changed in the Bailey shop since work ceased. Filled with a variety of planers and saws, it smelled like a place where wood was still worked. Hanging from nails were belts and the wooden patterns for pump and separator parts. And scribbled or sketched on posts and walls were the dimensions of equipment and their designs. At a moment's notice, the marvelous and arcane tools of the cranberry grower could again have been made here.

No longer built, the cranberry separator survives nonetheless. The industry has yet to find a better way to part the good berries from the bad than with a machine invented more than a hundred years ago. So in the sprawling Ocean Spray plant in Middleboro, the bulk of the Massachusetts crop is processed in two ways. The majority of the berries are transformed into juice and sauce, cooked in gleaming cauldrons and routed through a network of glass tubing in a facility that is decidedly modern. Moving to the fresh-fruit packing room puts one back a good eighty years in time. In it stands an entire bank of varnished, middle-aged Baileys, turning out the fruit. It is a technology that cannot be improved on.

Screening cranberries, Ocean Spray Cranberries, Inc.
Lindy Gifford photograph

The ship *Savoia* arrives in New Bedford Harbor from the Cape Verde Islands, October 1914.
Courtesy of the New Bedford Whaling Museum

Chapter 5

Out of the Islands

In the many shops on Cape Cod that cater to the summer tourist, it is still possible to buy postcards of the scoop harvest. Although the mechanical picker had killed manual picking with cranberry scoops nearly everywhere by 1960, the image of the old harvest, a Cape Cod tradition, remains. In colors always far too bright, dark-skinned men and women are shown bent to their work or carrying boxes across a green bog. On the back of the postcard, the caption reads, "Cranberry Pickers, Cape Cod, Mass." One wonders whether the recipients of these postcards, and many of the senders, know the identity of the pickers. Do they stare at the picture and ask themselves who these people can be?

The answer would surprise them, for the most obvious choices—a remnant group of Native Americans, or black migrant workers from the rural south—are wrong. These people are Cape Verdeans, and their first home was a distant archipelago in the Atlantic Ocean about 400 miles off Africa's western coast. Almost all that has been written about Cape Verdeans and their part in the life of the region is confined to the postcard caption: they picked cranberries. As such, Cape Verdeans in southeastern Massachusetts live a postcard life, their identity and history seemingly limited to having been cranberry pickers. In truth, the immigration of Cape Verdeans to America and their lives here are a rich and fascinating story.

Frequent contact between America and the Cape Verde Islands began in the eighteenth century, when East Coast whalers hunting in the southern Atlantic put in to these islands to replenish stores or seek new crewmen, who accepted smaller wages than American sailors and were far more tractable. The pod of thirteen tiny volcanic islands had long been a refuge, far from any continent, for traders, smugglers, pirates, and vagabonds. Colonized by the Portuguese in the sixteenth century, the islands were a stopping point for slavers bringing captured human cargo from West Africa to the New World. Out of the cultural intermingling of Portuguese and West Africans evolved a language distinctive to the Cape Verde Islands: Crioulo. Out of the physical intermingling—the liaisons between Portuguese colonizers, German, Italian, English, and French sailors, and West African women—came the Cape Verdeans, a people both exotic and handsome.

By the beginning of the nineteenth century, a small but persistent link existed between the islands and New England. More and more Cape

Verdeans could be found on American whalers, having gained a reputation as exceptionally skilled seamen. In Nantucket, the largest whaling port of that era, a third of all crewmen were Cape Verdean. Those who left their home islands for work on a whaler made little money but gained much in opportunity by leaving the Cape Verdes.

On the arid, mountainous, leeward islands of Brava and Fogo from which many departed, most Cape Verdeans lived in poverty as tenants, working the land of large farmers. And periodically, life in the Cape Verdes changed from poverty to extreme misery. The misery was brought by the prevailing southeast winds, carrying the hot and dry air of Africa's Sahara Desert to the islands. In some years, those winds would bring no moisture with them, and the Cape Verdes would suffer. Between the years 1830 and 1833, drought and the disease and starvation it brought killed 35 percent of those who lived on the islands. Thirty years later, drought reigned for another three years and 40 percent of all Cape Verdeans died. It was a cycle that could not be stopped and would continue with mechanical regularity in the next century.

During the last decades of the nineteenth century, the Yankee populace of coastal Massachusetts became aware of the presence of these dark foreigners in their towns. In the dying whale industry, where oil from the whale had been eclipsed by oil from the ground, Cape Verdeans were trimming more sails and cutting more blubber. When Yankee captains finally sold the ships that had brought so many stunning fortunes, Cape Verdean mariners bought them up, gathered a crew, and continued to whale for whatever could still be made on whale

Immigrant laborers at work on a new bog.
Middleborough Public Library

oil and whalebone. And when they tired of the two-year voyages and a transient life, many Cape Verdean men did not return to the islands and poverty, but came ashore in cities like New Bedford and found work along the waterfront. In the rope-walks and cooper shops, Cape Verdeans took jobs where their maritime skills served them well. At the docks, they rigged vessels and labored as long-shoremen. On the sovereign whaler, where crew-men of many backgrounds came together, skill was the basis for judging a man—how agile he was in the rigging, how steady his hand and eye when har-pooning. In hunting whales on the open sea, Cape Verdeans were beneath no other men. But once on shore, Cape Verdeans lost a right they were never denied on a whaling ship: equality. Living and working in the community of white New Englan-ders, Cape Verdeans immediately became lesser beings, easily denigrated and scorned.

The sea skills that a small number of Cape Verdeans perfected on American ships in the early and middle years of the 1800s served many Cape Verdeans at the end of the century. For it was then that several circumstances coincided to bring large numbers of people from the Cape Verde Islands to southeastern New England. In 1885 1,347 acres of cranberry bog existed in Plymouth County, the new center of the industry that seemed to offer unlim-ited swampland for bog building. As splendid as this growth was, it precipitated a grave problem for cranberry growers: they no longer could find a suf-ficient number of workers to harvest the crop. As acreage was expanded and cranberry companies grew, the harvest became a much larger, less per-sonal event. The snap machine was in use, and in some places the cranberry scoop. Both tools made picking more efficient, but in the minds of many, more tiring and difficult as well. What had been an Indian Summer idyll on Cape Cod, a carefree social time when fun was had and money was made, now became agricultural drudgery in Plymouth County. Yankees continued the harvest tradition on the Cape's small, locally owned bogs, but were unwilling to labor on the big bogs where cranberry growing had become such a business.

So the need arose for new laborers in the cran-berry region, men and women who could not choose their work and would accept whatever jobs were offered. These laborers were foreign-born immigrants whose arrival in America coincided with the great expansion of cranberry growing. Finns, Poles, Italians, and some Irish made their way south from Boston to Plymouth County. Some came on their own, having heard of seasonal work to be had, others were sent out by employment agencies, and the rest were herded there by overseers or padrones of the same race, who gained the loyalty of immigrants only because they were fearful of getting along in a new land with new ways. For the most part, the Europeans did not work out. The accepted view of the day, and for years to come, was that entire nationalities of foreign birth could be characterized by varied personal qualities. Yankee growers found the Irish hot-headed, the south Italians quarrelsome and hard to please. The Finnish were considered excellent at labor like bog building, ditching, and sanding, but picked rather slowly. It was said that Poles were fine workers with an even temperament, but disliked the seasonal work. They sought steady employment elsewhere. In Plymouth County and on the Cape, only the Finns would remain to play a part in the cranberry industry.

Cape Verdean mariners living and working along the southeastern Massachusetts coast saw the labor shortage on the bogs in another light: a chance for many Cape Verdeans to leave the islands and pick cranberries in America, and for themselves, an opportunity to make money transporting them here. In order to do so, Cape Verdeans purchased the dregs of the New England coasting fleet, schooners that had been worked hard hauling coal, lumber, or ice. There were always a few available for a little money, ships so old and derelict that even the tightest Yankee captains had no use for them. These were the vessels Cape Verdeans could afford to buy. In the 1890s, the voyages between the islands and New England began. Collectively, they were known as the Brava Packet Trade, a reference to the small island most ships sailed to and the seasonal regularity with which they went.

A voyage might begin in spring or early summer from a harbor in Brava, where a two- or three-masted schooner built in Maine or Massachusetts and now flying the Portuguese flag, lay. As few as fifteen, or as many as one hundred men, women, and children boarded her, each assessed fifteen dollars—twenty to thirty dollars in later years—for the crossing. With passengers, crew, and often no other navigational aid than a quadrant, a captain set out in a ship leaky and nearly on its deathbed. With fair winds and weather, the voyage to New Bedford or Providence was made in a month or less. But as often as not, fate intervened in the form of a gale or a hurricane and masts were toppled, the rudder smashed, the vessel blown far off course. A Brava Packet Trade captain, accustomed to treading the thin line between success and disaster, would jury-rig new masts, repair the rudder, and sail on. But now the trip would lengthen to fifty days or more; those on board waited out storms in sickness below deck, and helplessly watched days pass when the vessel was becalmed. Provisions took the form of dried foods such as corn and beans, which could be stewed into the traditional Cape Verdean dish, manchupe. Steamed semolina became the nourishing standby called cous-cous. These stores would only last if the journey was no more than moderate in length. For the passengers, the crossing was the first of many trials they would know in establishing themselves in America.

The schooners from Cape Verde, latter day *Mayflowers* with their own pilgrims, docked in summer in either New Bedford or Providence with their load of uncertain and fearful passengers, who ran straight into the arms of waiting relatives. Immigration officials classed the Cape Verdeans, who arrived with Portuguese passports, in one of two ways. Those with light-colored skin were

Cape Verdean immigrants aboard the *Savoia.*
Courtesy of the New Bedford Whaling Museum

recorded as white Portuguese and officially grouped with immigrants who entered America from Portugal's other island colony, the Azores. Cape Verdeans whose skin color was deemed more African than European were noted as black Portuguese or simply Bravas. In reality, the distinction between white and black Portuguese was a general distinction between Azoreans and Cape Verdeans. From the start, the class difference that Yankees perceived in skin color created racial tension between the two groups. Azoreans, eager to be accepted by the region's natives, denied any association with the Cape Verdeans.

With passengers departed, the schooners waited out summer's remaining days and the weeks of early fall when the cranberry harvest took place. All the while they were filled with goods for the return trip to the Islands: lumber and other building materials, clothing, shoes, and various manufactured goods. In the drought years, hundreds of pounds of bread and flour were loaded on board for a people to whom extreme scarcity was a way of life. By mid-October, the last of the cranberry crop was picked, and at the wharves, final preparations were made for the trip back. Cape Verdeans now settled

Harvesters and overseer.
Courtesy of the New Bedford Whaling Museum

in America booked passage for a visit home; some who had come for the harvest returned at its conclusion with money that made them rich by island standards. Many carried parcels, letters, envelopes fat with dollars, and less tangible but deeply felt greetings sent by Cape Verdeans in America who could not return to family and relations.

Undoubtedly, there were many different routes from New England to the Cape Verde Islands, depending on the captain's preference and the weather that his ship faced, but one route existed which allowed some contact with land. Its first leg took a schooner 700 miles in a southeasterly direction to Bermuda. Then came the longest stretch of all, 2,000 lonely miles northeasterly to the Azores. From there the passage to the Cape Verdes was 1,400 miles south and a little east. Winter always found a few Brava packets trading the islands' primary export, salt, along the African coast, or carrying passengers to the West African port cities of Dakar and Bissau. But by springtime, the schooners were back at their home port, their captains preparing for the long trip to Massachusetts and cranberry picking.

It was in the years between 1900 and the 1920s that the Brava Packet Trade reached its height. In that span of time, a dozen ships often made the annual trip. There were still the best of reasons to leave the Cape Verdes: drought and death. In 1902 and 1903, a quarter of all islanders succumbed to the effects of unrelenting, killing weather, and again in the early 1920s, nearly that many Cape Verdeans died of the same causes. It was tragedy on a scale Americans can barely imagine.

This desperation at home made passenger smuggling a common occurrence on packets crossing the Atlantic early in the century. After Portuguese officials had checked manifests and passenger lists, extra passengers were furtively brought on board. On occasion, when a vessel run aground on New England's coast required a tow, the Coast Guard would find this extra cargo, hidden in the depths of the bilges. But more often they slipped ashore by dory at night in the harbor of Providence or New Bedford, before the boat docked and immigration officials were met. To the packet captains, smuggling meant money easily made; to many Cape Verdeans who had entered America legally, it was shameful, a stain on the character of a people considered highly moral.

There was yet another legal wrinkle in the immigration of Cape Verdeans to New England, and it involved cranberry growers. A number of growers, concerned about finding sufficient pickers for the harvest, are thought to have contracted with captains to bring a specified number of men from the islands. From the wages of the newly arrived pickers, the cost of passage was deducted. Since 1885, however, the importation of contract labor had been forbidden by legislation known as the Foran Act. The goal of the law was to save American jobs for Americans, but it rang hollow on the cranberry bogs of Massachusetts, where few but Cape Verdeans would do the work. No evidence of contracting between a captain and a cranberry grower was ever discovered—not a bad thing, considering the special circumstances of the cranberry industry.

Through the summer and fall of each year, the Brava Packet Trade continued, bringing perhaps 1,000 men, women, and children to New England's shores every year. By 1912 there may have been as many as 15,000 Cape Verdeans in America. The size of the population is hard to measure accurately, since immigration officials classed Cape Verdeans by color, either white or black Portuguese, and not by their place of origin. What eventually curbed the influx was American law, which was amended in the early 1920s, limiting immigration by "people of color." Between then and 1965, when immigration laws were again expanded to encompass the many, fewer Cape Verdeans came to live in America.

A small number of Cape Verdean mariners kept up the trade, though the vessels, at the end of their useful lives, seemed to change yearly. They were schooners like the three-masted *Charles L. Jeffrey,* built in 1881 and lost in the Cape Verde Islands in 1924, and the *Frank Brainerd* of Rockland, Maine, abandoned in the mid-1930s halfway between Providence and the islands, all on board rescued by the Coast Guard. One Brava Packet captain owned in succession as many as thirty different vessels during his time in the trade.

The 1930s were a tragic decade for the Brava Packet Trade, with five ships lost in the ten-year period. In the fall of 1934, two packets returning to the Cape Verdes went down in autumn storms. The *Manta* carried a crew of nineteen and thirteen passengers, but it is not known how many lives were lost when the *Winnepesauke* foundered in the same seas.

The Atlantic crossings ceased during World War II and with them, communication between islanders and their relations in North America. Once the conflict subsided, the traditional link, now more than fifty years old, was reestablished.

The three-master *Lucy Evelyn,* which had carried potatoes and lumber along the New England coast, was brought into the trade, as was the former Coast Guard vessel *Illgria,* newly christened the *Madalan.* Perhaps best known of the postwar packets was the *Ernestina,* which, as the *Effie Morissey,* had brought scientists and explorers above the arctic circle. By far, these ships were the safest, best-equipped packets ever to sail in the trade.

In the era after the war, when technology was transforming nearly every aspect of American life, even New England Cape Verdeans were doubtful whether sailing ships, by now considered outdated, should be making the long ocean voyages. Yet, as before the war, the packets remained the only cheap and direct means to the islands. Another cycle of drought and widespread death in the late 1940s served to convince Cape Verdeans that, as a means to send desperately needed food and supplies back home, the Brava Packet Trade still served a purpose. So it did, for another two decades. One by one, the remaining packets dropped out of the trade. The *Ernestina* made her last passenger crossing from the Cape Verde Islands to Providence, Rhode Island, in 1965, her owner and captain, Henrique Mendes, then eighty-six years old.

The *Ernestina* can again be seen in southern New England, a gift from the government of Cape Verde to the United States in 1982. Out of New Bedford Harbor, the *Ernestina* sails again, but this time to coastal cities and towns where Cape Verdeans make their homes, a living reminder of their heritage and of the struggle to come to America and start life anew.

The *Ernestina* docked at Onset, Massachusetts.
Lindy Gifford photograph

Children scooping, Wareham, Massachusetts. Lewis Hines photograph
Library of Congress, Prints & Photographs Division, National Child Labor Committee Collection

Chapter 6

Living A Postcard

During the first decade of the 1900s, the U.S. Immigration Commission undertook a lengthy investigation into the lives of the foreign-born who were changing, so quickly, the social fabric of American life. The Immigration Commission sent its men among immigrants working in industry and agriculture. The agents looked into homes and lodging, examined the foods immigrants ate, and observed their dress. They spoke with bosses, townspeople, and bankers—seemingly everyone except the foreigners themselves. When the Immigration Commission came to southeastern Massachusetts it was not to investigate the Finns, but to study the habits of the cranberry pickers, then known as Bravas or black Portuguese, who had so recently arrived in large numbers. The commission wrote in 1911:

> They are rapidly taking upon themselves all the unskilled work connected with the preparation, the planting, the cultivation, the care, and the picking of the bogs. Twenty years ago there was scarcely a black laborer in the cranberry district; ten years ago they were beginning to come in earnest; today they are driving before them the last of the Americans, the Poles, the Italians, and the Finns, and are proving themselves the best pickers and the best wheelbarrow men who ever came upon the bogs of Cape Cod. Hundreds of them are employed the year round, and something like three thousand are employed during the harvest.

The Immigration Commission's description of Cape Verdeans settling in Massachusetts made it sound like an invasion of dark faces and bodies; undoubtedly, some Yankees must have thought that it was.

The standards the Immigration Commission used to judge the Cape Verdeans were the standards it used to judge everyone: those of white, middle-class Americans. Anthropology was still a very new discipline, and the idea of studying another people's social habits and customs in an attempt to understand them was not widely known. Expected to measure up to their Yankee neighbors' ideals, there was no way that Cape Verdeans could hope to be understood. The commission would learn, in its own way, what living a postcard life was about.

The Cape Verdean immigrants being observed by the government at that early date were mostly single men who had come to work a few seasons in America, intending to return home with money to buy land and live well in the Cape Verde Islands. Others were working to save enough to buy passage for their wives and children to New England. Only a small number had brought their families with them. The towns in which the Immigration Commission found the Cape Verdeans are the same ones in which many of their children and grandchildren live today: coastal towns like Harwich, Wareham, Plymouth, New Bedford, and Providence. Most stayed close to the cities where they had disembarked from the islands and close to the source of their staple fall employment, the cranberry bogs.

Newly arrived, unable to speak English, and black, Cape Verdeans were inevitably at the absolute bottom of the region's social and economic scale. They were bound to start out with the meanest housing, the poorest food, and subject to the worst prejudices.

> Those who live in their own homes in the small towns or in the woods near the bogs find shelter in the poorest of lodgings. Old abandoned houses, cheap shed-like structures, or rooms in some tenement house where each family or gang eats and sleeps, are usual. In one instance thirteen men lived in a two-story building about sixteen by twenty feet, some little distance from the bog. They slept upstairs on straw or cheap mattresses and cooked and ate on the ground floor. One of the number was appointed each day to do the cooking. The whole place was very bare and dirty; filth and litter of all sorts abandoned and evidently no pretense of cleaning was made.

Once Cape Verdeans made up the majority of the cranberry harvesters, growers put up bog houses on their land for those who had no permanent housing or had come out from New Bedford for the harvest. But these lodgings were no better than what Cape Verdeans found on their own, for they were built to suit the Yankees' impression of their workers.

Some of the shacks provided are of the roughest and rudest sort imaginable; mere one-story sheds about eight by ten feet, rough boarded, with a shed roof of rough boards. The bunks are rude boxes filled with straw. That anything will do for six weeks is the idea.

—U.S. Immigration Commission, *Immigrants in Industries*

> On other bogs ten-by-twelve-foot two-story houses are built to accommodate the transients. There are a stove, benches and a cheap table below, and a couple of four-foot bunks of rough lumber above. Four or five or more occupy each of these houses. These are cleaned periodically under the supervision of the superintendent, but otherwise no attention is paid to sanitation. The rent of these houses is not much; sometimes six to eight dollars a month, sometimes nothing. Some of the shacks provided are of the roughest and rudest sort imaginable; mere one-story sheds about eight by ten feet, rough boarded, with a shed roof of rough boards. The bunks are rude boxes filled with straw. That anything will do for six weeks is the idea.

Cape Verdeans called these buildings—never really substantial enough to be dignified as bog houses—"shanties." While some were occupied only at harvest time, others were meager shelter nearly year-round for men who came in the early spring and worked on the bogs until winter. In the big empty stretches of Massachusetts pine forest and bog, some shanties still stand on the land, decrepit and abandoned. Lonely-looking sheds in

Cape Verdean and other bog workers at lunch. Ocean Spray Cranberries, Inc.

Cranberry worker housing, Wareham, Massachusetts. Lewis Hines photograph
Library of Congress, Prints & Photographs Division, National Child Labor Committee Collection

A bog shanty.
Lindy Gifford photograph

isolated places, they are a reminder for the growers of times past and for the Cape Verdean Americans, a glimpse back to the first lives and jobs of their people here.

Inside workers' shanties, the Immigration Commission investigated cooking pots and the Cape Verdean diet. Finding little variety in the dishes prepared and eaten day after day, they quickly assumed ignorance on the part of the men.

When they first come to the bogs, they know very little about cooked food and almost nothing about preparing it. They actually suffer want until someone teaches them the method of getting and preparing food. They eat rice, beans, and pork, and use a great deal of lard. Some biscuits of white flour are made for the evening meal, the heartiest meal of the day, but milk crackers and bakers' cakes serve at the other meals. When they first came to one large bog ten years ago, their menu consisted of ripe maize or Indian corn on the cob. This they boiled until partly softened and ate with molasses poured over it. But this diet has now given way to a much better variety of food. The butcher's wagon and the baker's cart make regular trips to bogs when many are employed, and the grocery stores send out quantities of canned goods, canned roast beef, flour, and beans to some of the gangs. Those who remain add sweet corn, potatoes, and string beans. They are living better each year, but their cooking is not good, and there is a tendency to depend largely on crackers and bakers' bread. The diet is more largely vegetable than animal, but the cost of living is somewhat higher than that of the Italians, and fully up to that of the Finns. The food both in variety and preparation grows gradually better. Lima beans, which retail at twelve cents a quart, and rice are staple articles of diet for the Portuguese throughout the cranberry section.

Scoopers. Middleborough Public Library

Lima beans and rice is still a favorite dish of older Cape Verdeans; bags of each are often seen in their shopping carts at local supermarkets. Beans and rice is also a traditional Cape Verdean dish, which explains their importance in the diet of cranberry bog workers in the early 1900s. It was for these reasons—familiarity and tradition—that the earliest immigrants ate so much corn. In the islands, corn was a staple of every diet and a part of many dishes. But these explanations for Cape Verdean eating habits were ones the Immigration Commission, in its ethnocentrism, would never have found. Other facts explain the modest diet of the early Cape Verdeans. The simple fact was never acknowledged that after a day of bog work, an exhausted body prepares the simplest thing, nor that diet is a function of income, and Cape Verdeans, as we shall see, took great pains to save as much as possible of their slender wages. And no doubt it took time for these men to become familiar with new fruits and vegetables grown in New England.

The first years of immigration were the very hardest years for Cape Verdeans in America. Early photographs of men and women arriving in New Bedford from the islands are telling. The fear on their faces is palpable, a fear of things different and a new land, language, and customs, all entirely alien to what they knew. Unused to the climate and the food, living in miserable and overcrowded conditions, Cape Verdeans died, as a Falmouth doctor wrote, "like flies," from tuberculosis and pneumonia.

It was the lure of work on the cranberry bogs that brought Cape Verdeans to America in numbers, and for two generations the industry provided them with a stable income in the autumn months. Picking season also brought Cape Verdeans the highest wages of all the jobs they held in a year, but not without a price: it was the hardest work, as well. Pickers worked between the hours that the bog dried and the dew fell again, generally from six and a half to seven hours a day. The nature of the work, the position it forced the body into—bent over and kneeling—more than made up for the modest daily hours. The common scene was a line of men maybe twenty in number, on their knees and pushing scoops ahead of them. They pulled along field boxes into which the fruit was emptied, each holding thirty-three pounds of cranberries when full. These were carried to the bog edge or shore by the pickers or removed by workmen using wheelbarrows. Overseeing all was the boss picker, sometimes a Cape Verdean, who made certain that the picking was clean.

For an hour's scooping, the worker made thirty or thirty-five cents, realizing at least two dollars for the day. When the scooping was done on piecework, the stakes were higher and some men brought home five dollars or more. But piecework was not universally approved of, for many growers thought it made the pickers hurry and leave too many berries on the vines. Where hand picking was still practiced, eight cents was paid for each six-quart tin measure picked. Nimble-fingered women tended to do this work, which brought them about a dollar and a half for the day, the lowest wages on the bog. Snappers, with their tiny mechanical scoops used on young vines, worked at the same rate as hand pickers, but could gather twice as many berries. Consequently, wages from snapping were often the very highest earned in a day.

So the work went, in the harvest that stretched from the beginning of September to mid-October, with rain days the only break in the work. Cranberry pickers were generally not paid until the end of the season, a tradition that extended back to the days when Yankees picked the crop, and many had to get along on savings or advances from the grower until the final day came. When it did, many had earned seventy to eighty dollars for six weeks of work. For unskilled work, this was considered a good wage. What Cape Verdeans did with their earnings must have astonished the Immigration Commission, for it was far outside its general perception of black workers.

> They begin to save as soon as they begin to earn. The pickers, when paid weekly or biweekly, go at once to the nearest town and deposit their checks, or, if at some distance, appoint one of their number each pay day to make the trip. In Harwich, Plymouth, New Bedford, Wareham, and other places certain savings banks have hundreds of black Portuguese depositors. The savings bank in one of these villages in Massachusetts, in the cranberry district, is illustrative. This bank has approximately five hundred black Portuguese depositors, with accounts averaging about two hundred dollars to three hundred dollars each. These accounts run from two to six years, and the individual credits foot up eight hundred dollars, one thousand dollars, or even fifteen hundred dollars in some instances. The bulk of the deposits are made in the early autumn, and come in the shape of pay checks, but there are many who deposit all through the summer. "About October fifteenth," said the bank president, "they draw out their deposits. In 1907 we paid out about fifteen thousand dollars to black Portuguese at that time; in 1908, the amount drawn was over twenty thousand dollars; this year it will probably exceed either 1907 or 1908." In another village, a five-cent savings bank has one hundred black Portuguese depositors, and no account exceeds one hundred dollars. The disposition to save is universally commented on; and perhaps a larger proportion of Portuguese than any other race doing unskilled labor make use of savings banks. There are some spendthrifts, but on the whole frugality and the disposition to save are characteristic. Probably fifty percent of the savings are sent to the old country, a small percentage for investment, but the most for the support of destitute relatives, or to bring members of the family to the United States.

The sum of money received at the harvest's end served another purpose the Immigration Commission did not know about: it helped Cape Verdeans through the long, slow Massachusetts winter. Some harvesters returned on the packets to spend the winter among family in the islands. Others who had come out from New Bedford or Providence returned to find work in the textile mills or

Selling Mayflowers at the Harwich, Massachusetts, station.
Harwich Historical Society

on the docks. But for those who made their homes near the bogs in Plymouth or Wareham or Falmouth, winter work was scarce—cutting wood or cutting ice on the ponds was about all that existed. Harvest wages spent carefully on food and necessities helped people along until spring again brought warmth and work to the region.

The month of April found some Cape Verdeans back on the bogs, perhaps clearing the ditches of mud and growth or setting out new vines, but others began a round of harvests that would not end until the cranberry crop was once again in. In earliest springtime, they entered the pine woods to pick the tiny pink- and white-petaled trailing arbutus or Mayflower. Made up in bunches, the flowers were sold to buyers for the Boston market or in the train stations of the towns. Come June the strawberries in Falmouth's acres of beds were

About October fifteenth, they draw out their deposits. In 1907 we paid out about fifteen thousand dollars to black Portuguese at that time; in 1908, the amount drawn was over twenty thousand dollars; this year it will probably exceed either 1907 or 1908.

—president of a Massachusetts bank, *Immigrants in Industries*

Children scooping. Lewis Hines photograph
Library of Congress, Prints & Photographs Division, National Child Labor Committee Collection

Cape Verdean children at school, Harwich, Massachusetts.
Harwich Historical Society

ripe and that harvest provided a few weeks' work. Later in summer the region's sandy soils offered up another crop found in grown-up fields and woodlands: wild blueberries. Cape Verdeans picked them for fifteen cents a quart, sometimes picking as many as ten quarts in a day. As August gave way to Indian Summer with its mellow days and chill nights, men, women, and sometimes children waited for the word that cranberry picking would begin. Resourceful and attuned to the seasons, early Cape Verdean immigrants moved through the year harvesting flowers and fruit in a seasonal round. Filling in the gaps with whatever came their way and living frugally, they made do.

As Cape Verdeans became established in southern New England and slowly adapted to the landscape and culture, they became more prominent in the white world. And while their heightened presence on town and city streets and in shops meant that they were settling in, it also forced greater contact between themselves and whites, contact that the white community found unpleasant. The commission reported:

> There is noticeable a growing sense of their importance. Electric car conductors, trainmen, and others have become well aware of this fact. Some of the Portuguese are beginning to have "rights." A few years ago it was easily possible to put a Brava on the back seat in a street car and to make the "jim-crow" car idea a practicable expedient. Now the Brava who knows his importance refuses to move back or forward or anywhere else until he pleases to do so, much to the annoyance of the conductor. Said an old electric trainman: "They are making themselves more offensive to us on the cars and to the public generally every year." Frequently the bolder spirits seem to take delight in sitting down in the same seat as a white woman, if there is an opportunity. The white patrons complain, but there is no legal method of putting the Brava into any seat he does not choose to occupy. This independent attitude is a matter of a very few years, but it increases steadily.

Yes, I remember the first ones. We thought they were wonderful, the children that came to school. If we could sit beside the black child, oh, that was wonderful.

—Eunice Bailey and Jennie Shaw

What the Immigration Commission interpreted as a growing sense of self-importance among Cape Verdeans was no more than their claiming the same rights they saw exercised daily by people all around them. While white New Englanders could not legally disenfranchise the Cape Verdeans in their midst, they could exert the sort of social pressure that ethnic majorities have always used to keep ethnic minorities in line. This they did and would continue to do for years to come.

Despite the friction between whites and Cape Verdeans, their children attended school side by side, although there were occasional calls to segregate the schools. The racism that pervaded adult relations did not extend down to the youngsters, who regarded each other with a mixture of curiosity and delight. An elderly Carver woman recalled, "Yes, I remember the first ones. We thought they were wonderful, the children that came to school. If we could sit beside the black child, oh, that was wonderful."

If the Cape Verdeans were becoming increasingly "independent," at least they were moral, a quality the Immigration Commission seldom ascribed to foreign populations in America.

Reliable under supervision, docile, obedient, willing to work, and not over-fastidious with regard to food or shelter or the discomforts of the weather, and apparently satisfied with the isolation and the somewhat disagreeable work, they are very desirable men all the year round in the cranberry district.

—U.S. Immigration Commission, *Immigrants in Industries*

The Bravas have a much higher moral code than one would suppose, judging by their ignorance and standard of life. While they are a social people who love music, dancing, and frolics, they are generally temperate and the average of personal morality is fairly advanced. Rioting, drunkenness, and indecent revelings are said to be infrequent or (in places) unknown occurrences.

On the face of it, there was much in the Cape Verdean character that might have appealed to Yankees. After all, they were moral, temperate, and frugal, admirable qualities that New Englanders would quickly have used to describe themselves. But no affinity developed between the races, no grudging respect among the Yankees for the Cape Verdeans. The black man's foreignness was too great, and the matter of color created a void that could not be bridged. The Immigration Commission summed up the Cape Verdeans in this way:

Nearly all the Portuguese, white as well as black, are illiterate, but in addition to their illiteracy the Bravas are stupid. Of the many who deposit in savings banks, very few indeed can write their names. They are ignorant day laborers only, and as such fill a much needed place in the supply of seasonal labor on the Cape. Employers agree in praising their efficiency as pickers and their general work as unskilled laborers. Reliable under supervision, docile, obedient, willing to work, and not over-fastidious with regard to food or shelter or the discomforts of the weather, and apparently satisfied with the isolation and the somewhat disagreeable work, they are very desirable men all the year round in the cranberry district; furthermore they are almost the only men who can be obtained in sufficient numbers to supply the demand for bog laborers.

Even among recently arrived immigrants in southeastern Massachusetts, Cape Verdeans were at the absolute bottom of the pecking order. The white Portuguese, who had come from the Azores and the mainland in large numbers to work in New Bedford and Fall River's cotton mills, denied any relationship between themselves and the Cape Verdeans. "Many have a strain of white blood, although the white Portuguese repudiate the idea that there is any blood relation between them and the blacks," reported the Immigration Commission. The Portuguese made this claim despite the fact that Portugal itself had colonized the Cape Verde Islands and imported West African slaves to work on the sugar plantations.

Relations between Cape Verdeans and the Finns, the only other immigrant group to remain in the cranberry industry, were no more equitable. The Immigration Commission provided a glimpse of this in recording the following:

Both Finns and Italians are said to possess more intelligence, and when both Finns and Portuguese are employed on the same job, the Finns refuse to perform certain tasks. Thus the Finns will not wheel sand to the bog. When no Portuguese are employed, no objection is made and the Finns handle a wheelbarrow without question.

Owner, overseer, Cape Verdean harvester, Westport, Massachusetts.
Courtesy of the New Bedford Whaling Museum

As white Europeans, Finns had no more use for the Cape Verdeans than did native New Englanders. Having shunned eastern and southern Europeans on the passage to America, their response to blacks was predictable.

In the 1930s, a novel came out of the Massachusetts cranberry country that described, more honestly than any other source, work on the bogs and relations between blacks and whites in the era. It was called *Cranberry Red* and was written by a Plymouth native named Edward Garside when he was in his early twenties. Garside's book follows for a time the life of a young man, sensitive and college educated, who, drifting from job to job at the end of the Great Depression, lands at a cranberry cannery and the bogs in a small Cape Cod town. To some degree, the story was Edward Garside's, for he worked on the bogs in his hometown to earn money for college and to support an aging father.

Garside captured perfectly the aura of the cranberry region, the stark vistas of bogs and pitch pines, the odor of wood from the buildings, a certain poverty in the landscape that runs over into the small towns, insular and provincial. He rendered the work in a way that was ironic, gritty, and true.

Lunch break at United Cape Cod Cranberry Company.
Ocean Spray Cranberries, Inc.

Digging ditches through heavy muck and alder roots with the feet freezing in rubber boots was not the pleasantest kind of labor.... But sanding was the worst. It was a nightmare, a penance, a scourge. And for two meager dollars, two lousy bucks, two, sad, pathetic, lonely, slugs, eight bits, a day.

During the picking season, matters were only slightly different. Rainy days, if there were not too many of them, were welcome occurrences. There were so many strangers about. One could assume a proprietary air, if one were regularly attached to a bog and not just hired for the scooping. There was news from the [Cape Verde] Islands, there were strange women to look at, and to get to know. There was no telling what might happen. Once in a while there was a murder or near murder, after a drinking bout, an exciting stabbing or cutting. And most important of all, there was money in the air. Those were gala days.

Cranberry Red made a strong impression in the growing region when it was published, but not because the cranberry region was the backdrop for the novel. The book caused a stir, in the words of its author Edward Garside, "because, I suppose, [of] the somewhat realistic treatment between the Cape Verdeans and the owners and bosses. These things you didn't talk about then, you know." But Garside had chosen to talk about them, and paraded the fear, suspicion, and loathing between the races across four hundred pages. A Cape Verdean is crushed to death in a cannery accident, the police routinely break up a Cape Verdean dance and do their utmost to keep the blacks in line. Cape Verdeans threaten to torch the screen-house of a tough, embittered grower. In events such as these, the novel continually points up the thinly veiled hatred felt by blacks and whites alike and their total inability to understand each other or to care to.

Garside's protagonist, young Keith Bain, views these events thoughtfully, leaving no doubt as to where his sympathies lie. But interestingly, Bain's response belies a sense of ambivalence about black foreigners that even compassionate whites of the period apparently felt. Bain says of the Cape Verdeans at one point, "These people are brutalized. They're poor. You understand? They're poor, syphilitic, ignorant, and run into the ground." His compassion for the cranberry workers is evident in many places throughout the book, but it is always a distant compassion, mixed with a certain abhorrence of their ignorance, their looks, their color.

He forgot the man's blackness, the soulless eyes the color of a pointer's and the odor of negro sweat dried into the man's clothes. Whether it was because of his own fatigue, or because the black Portygee's simple, but impenetrable, defenses had crumbled away, he saw the man.

Digging ditches through heavy muck and alder roots with the feet freezing in rubber boots was not the pleasantest kind of labor.... But sanding was the worst. It was a nightmare, a penance, a scourge. And for two meager dollars, two lousy bucks, two, sad, pathetic, lonely, slugs, eight bits, a day.

—*Cranberry Red*

He was neither chicory nor cocoa in color, but fairly close to cafe au lait. His set face was mute as a soap box, was quite unrefined by the Portuguese blood in his veins. Like a small, stooped monkey he poked around, trying hard, but not getting much done. His heavy jaw hung forward, drawing down the hollow cheeks, liberally pitted with small pocks, and the bluish lips, like cuts of frozen veal, were forever pursed neatly, gently, just touching, as if they were about to bestow a trembling kiss of gratitude on the world at large.

A forgotten people, a people worked to the bone, Cape Verdeans were also to Bain utterly incomprehensible. "They were the victims not only of a steely Yankee greed and a silent will to live, but also of their own incapacity for adjustment to a way of living diametrically opposed to their own." Distant compassion may have been the only compassion that the most sensitive white New Englanders could feel toward Cape Verdeans, if Keith Bain is any example. In the 1930s and for years thereafter, racism toward dark-skinned people was learned at an early age in America. While one might pity the collective cranberry pickers in their plight, the sentiment toward an individual Cape Verdean was not strong.

I recognized that farming is very hard business. It depends on the weather and you can't expect people to be coddling you on a bloody cranberry bog. But on the other hand, there were inequities that were omnipresent, as a matter of fact, throughout the whole economy.

—Edward Garside

Edward Garside's social commentary encased in a novel was not lost on anyone, yet it was commentary that he himself felt the need to qualify.

I was trying to get a message across, of course, of the inequities of this world and they were very striking inequities, but not too strong. I recognized that farming is very hard business. It depends on the weather and you can't expect people to be coddling you on a bloody cranberry bog. But on the other hand, there were inequities that were omnipresent, as a matter of fact, throughout the whole economy. Yes I was aware of it, of course, but I wasn't a practicing radical or dissident, not at all, no. I remember talking to Mr. Atwood about it. He recognized that, too.

Ellis Atwood was a prominent Carver grower and a well-liked man. Reading *Cranberry Red,* he understood its message completely and told Edward Garside that he "thought it wasn't a bad representation."

In the autumn of 1933, a remarkable event took place on the Massachusetts cranberry bogs: Cape Verdean pickers went on strike. Left to their own devices. Cape Verdeans would have found it nearly impossible to organize the strike. As laborers, they were dependent on bog owners for work and wages, and as blacks and foreigners in a white land, they were powerless to change their own lot. But in June of that year, a labor organizer named J. J. McIntosh arrived on the scene. He was authorized by a man named Abraham Binns of New Bedford's Central Labor Union to sign up the Cape Cod cranberry pickers.

Whether motivated by a genuine desire to help black workers or a need to increase the number of dues-paying members, McIntosh struck a chord in many Cape Verdeans. A local paper, the *Wareham Courier,* speculated that the "Depression wages" being paid to workers for weeding and sanding were also fueling interest in the union. The Great Depression was still hovering over America, and business conditions were bad, but prices for cranberries were good compared to those for other produce. Labor and the workers thought that growers were not passing enough of the profits along to them.

At a June meeting in Wareham's village of Onset, two hundred pickers from Rochester, Carver, and Wareham were present and three hundred more signed cards pledging to join the union, known as the Cape Cod Cranberry Pickers Union, Local No. 1. The union quickly agreed on a schedule of wages that growers would have to meet:

Men scooping: 80 cents per hour
Women scooping: 70 cents per hour
Hand picking: 15 cents per 6-quart measure
Women screening: 40 cents per hour
Men weeding and sanding: 50 cents per hour for a five-day, 40-hour week

Also specified were a series of conditions. A workers' committee was to be allowed on bog properties to inspect living and working conditions. Pickers were to be paid at two-week intervals instead of at the end of the season. Within the two-week period, credit slips accepted at stores and banks would be issued. No child under seventeen was to be employed on the bogs.

An isolated bog shanty.
Ocean Spray Cranberries, Inc.

In 1910, when the U.S. Immigration Commission investigated Cape Verdeans, the cranberry grower was getting about $5.60 for every hundred-pound barrel of cranberries he sold. At that time, he paid scoopers thirty to thirty-five cents per hour and hand pickers eight cents per six-quart measure. In 1932 a barrel of cranberries brought the grower $7.80—an increase of more than two dollars a barrel—but hand pickers were still paid eight cents per measure and scoopers' wages increased by only five cents to forty cents per hour. Some cranberry growers, remembering the strike, said that the hourly wage issue was irrelevant, since most pickers worked on piecework. One recalled:

Every year they'd come back and it was all piecework. And a harvester in those days had thirty-five, thirty cents a harvest box, and he could pick on good picking, forty boxes a day. And in fact, I know one time on a real good heavy crop, some of them were averaging a hundred harvest boxes, and at thirty cents, that was thirty dollars [a day], and back in those days, you worked a week for twelve dollars. So you see, if you can make that type of money in the fall, you really were set for the winter, when maybe there wasn't any work.

Cape Verdean family on the harvest.
Ocean Spray Cranberries, Inc.

Many of influence feel the grower should be concerned with living conditions. Property owners have neglected to take an interest in the families living on the bogs and the outcome has been that welfare costs of the town have increased.

—New Bedford *Standard Times*

But in the year of the strike, T. F. Gonsalves, a Cape Verdean leader from Onset, declared, "As it stands now, pickers are paid by the hour when berries are thick and by the measure when they are thin."

As much as Cape Verdeans wanted higher wages, money was not the only goal of the union. Expressed in the written conditions that accompanied the schedule of wages was a plea for respect and concern from the cranberry growers and the community at large for workers and their families. It could be heard in the demand for inspections of living and working conditions, regular payment of wages, and an end to child labor. This last issue was a double-edged sword, however, since many Cape

Verdean families needed the money that children earned on the bogs. But the Massachusetts child labor law specified neither a minimum age at which children could work in agriculture, nor the number of hours they should work a day. For this reason, cranberry growing and Massachusetts farming and industry in general, were long enemies of the crusading National Child Labor Committee.

What the labor leaders wanted, they said, was a "humane society." The organizer McIntosh claimed that "exploitation of labor on bogs is worse than in any other part of the state. Sand carters are paid as low as two cents per load, sometimes carting sand for one quarter mile." Others declared, "Our people are living in unfit quarters at the edge of the bogs. Most workers are paying $3.00 a month rent and receiving 60 cents to $1.50 per day in wages." The cranberry growers disputed these wage claims made by the union, but it was harder to deny the statement regarding conditions in Wareham reported in the *New Bedford Standard Times:*

> Many of influence feel the grower should be concerned with living conditions. Property owners have neglected to take an interest in the families living on the bogs and the outcome has been that welfare costs of the town have increased.

As the harvest began in the first two weeks of September 1933, things began to heat up in Wareham and surrounding towns. Red and black banners declaring the new union's name were hung out in Onset. Striking workers wore armbands that said "PICKET" on their sleeves. Signs announcing "WE WANT A LIVING WAGE" were placed on trucks and driven through town. Strikers picketed a canning plant of Cranberry Canners, Inc., a growers' cooperative. By that time, from 600 to 700 had signed up as union members, still only a fifth of the number of pickers who normally worked on the harvest.

To cripple the harvest, the union would have to enlist far more numbers than it had to date. In an editorial, the *Wareham Courier*—always strongly supportive of its white middle-class patrons—asserted that the "supposed" labor organizer was gulling ignorant workers for their own profit and warning them that if they failed to join the union and tried to work on the bogs, they would be arrested. Each side in the struggle tried to play upon the uncertainty and confusion of the Cape Verdean cranberry pickers to its own advantage.

On September 8, the level of conflict rose when ninety-five workers at bogs in Manomet, Plymouth, refused to work for less than eighty cents an hour. An angry owner ordered them off his property. The mood was tense, threats were made, and an armed police guard watched over scabs who were brought in to pick the crop. Confrontations began to appear in more places; pickers walked off seven more bogs in Plymouth, and the strike spread to the lower Cape. There, 500 workers signed up with the union. In order to get their berries in, some Cape growers paid pickers from fifty to seventy-five cents an hour, an indication that the strikers' tactics were beginning to have some effect.

To really succeed, however, the union needed to paralyze the large bogs of Plymouth County where the majority of the Cape Verdeans lived and the bulk of the crop was grown. There, while some growers had faced constant disturbances, many others had gotten workers to pick for them without any difficulties. The greatest foe of the new union was not the cranberry grower, but the workers' need for money. The union could offer no one strike benefits, and the wages earned on the cranberry harvest were all that kept most families going through the winter. Despite the degradation of bog work, the harvest offered many Cape Verdeans something more tangible than the hopes of the Cape Cod Cranberry Pickers Union, Local No. 1.

In an attempt to force the growers' hand, the union switched to more heavy-handed tactics in the second week of September. As many as twenty carloads of young strikers were sent out into the bogs to call off pickers at work. The late Ernie Howes of Wareham, then a foreman for A. D. Makepeace Company, watched what took place.

I was a foreman at the Frog Foot Bog and I happened to look up one day and there was a whole parade of these, they were Cape Verdeans, most of them. And they were coming down the road and they were trying to frighten the pickers that we had.

The strategy usually worked, for it was often combined with threats to burn the picker's house and kill him if he did not leave the bog. These threats to property and person were extended to the grower as well.

The union employed a second tactic that, like the carloads of strikers, brought temporary results. It was well remembered by foremen and growers for its originality and impact. Once again, Ernie Howes was witness.

You know what they'd do? They'd have a meeting at night; these Cape Verdeans would have a meeting with these union men, and then the next day, they'd go right out picking, because they didn't want to lose anything. And so the union fella says, "We've got to stop that." So he went up to East Taunton and he hired King up there at the airport to fly this biplane down and he would dive down at those pickers to scare them off, and he did scare quite a lot of them off.

Gibby Beaton, who oversaw the picking on some of his uncle John J. Beaton's bogs, watched the biplane as well and wondered about the motives of some of the pickers.

I sometimes think they really weren't scared of that plane. They were more interested in getting up into the woods, hoping that the strike would benefit them. I mean, they kind of had fun at it, everybody'd laugh and joke and so forth.

He realized, as Ernie Howes had, that the Cape Verdean pickers wanted it both ways. They needed it both ways, too: some picking money to get through the winter, and better living conditions and higher year-round wages from the growers.

As the days passed, the situation became increasingly tense in the towns of Plymouth County and Cape Cod. In Carver, a grower was badly beaten, and a Cape Verdean man was shot through the hand in an altercation. Sixty-four strikers were arrested as a result, and heavy details of state police began to appear in the area. Vandalism occurred as well; strikers pulled up flume boards, flooding bogs, which then couldn't be picked, and several screenhouses were burned. In every town, men were deputized and carried sawed-off shotguns while patrolling the bogs in trucks. Some growers carried revolvers, and the men at A. D. Makepeace Company prepared themselves as Ernie Howes remembered:

We all had broom handles and we drilled out the broom handles and we poured babbitt [metal] in there, we were making these sticks, you know, night sticks. We never used them on the boys, but they...put up roadblocks to keep us from going to some of the bogs. It was quite scary at the time.

In the midst of the hysteria, when sides were clearly drawn—and whites were expected to stick together and blacks to do the same—a prominent Cape Codder leveled a surprising charge at Massachusetts cranberry growers. Chester A. Crocker, chairman of the board of selectmen in the town of Barnstable, wrote in a letter to the *Wareham Courier* that cranberry picking was grueling labor that required skill and rated an eighty- cents-per-hour wage. He went on to say that growers who paid

On the bog.
Ocean Spray Cranberries, Inc.

much less were motivated by greed or fear at having to keep up mortgage payments on bogs purchased at inflated prices. Growers would pay, he concluded, much more for damage done to bogs by poor pickers.

Crocker's letter had no effect on the situation, but seems today to have been an unusual, perhaps remarkable statement from an unexpected source. What did affect the strike was four days of torrential rain in the third week of September, which brought picking and attempts at picking to a complete halt. From this point onward, the strike seemed to lose momentum and falter. Efforts at arbitration between the strikers and growers failed, and the authorities took a step calculated to kill the unrest. Labor organizers J. J. McIntosh and Fred Woods were arrested for signing up workers under false pretenses. Released on bail, they were later convicted on the charge and given prison sentences of two months.

The September 29 issue of the *Wareham Courier* reported that the strike seemed to be dying; the pickers were returning to work and the growers were paying them higher wages, though not the eighty cents an hour they had demanded. By early October the Cape Cod cranberry pickers' strike had faded away. It was late in the season, and the Cape Verdeans needed to make some money for the months ahead. It is estimated that about 1,500 workers took part in the strike at one time or another, probably half the number that normally worked on the harvest.

Why did the strike fail? It failed partially because Cape Verdeans were confused, unsure why they should trust labor organizers any more than cranberry growers. It failed, too, precisely because Cape Verdeans were at the lowest rung on America's ladder. Uncertain of their place already, they were leery of taking such a drastic step against the white community. Yet the need to make money was more fundamental than any other reason. Without cash from the harvest, there would be very hard times ahead. The Cape Verdeans of southeastern Massachusetts would have to be better established and bankrolled before pushing for the right to live and work like other American citizens was possible. In the meantime, the cranberry workers returned to work assured of a little more money and of having taken an historic step. Their resistance on the cranberry bogs was the first strike by farm workers in Massachusetts's long history.

The desperate picture of life in the cranberry growing towns presented by *Cranberry Red* and the 1933 strike can be countered by recollections of growing up in the insular world of a Cape Verdean neighborhood and by some good memories of the cranberry harvest. Beatrice Pina lives in the coastal town of Marion in the house she shared with her late husband Vincent. On the same street lives Doris Gomes, Bea's half-sister. While each made some money from the cranberry in the past, more of their income has come from the waterfront estates at the other end of the road. At the grand houses owned by wealthy New England families, there are always lawns and flower beds to be cared for, meals to be cooked, and shirts to be ironed.

But these places could hardly be as cheerful as the neighborhood that Bea and Doris know. There, in good weather, old men and women sit on the porches and stoops of small wooden houses to watch the procession of cars, bicycles, and people pass by. In the streets, kids gather to talk and play games. It is a neighborhood alive. Sixty years ago the neighborhood was just as lively and people depended on caretaking, domestic work, and labor on the bogs to get along. But then, people lived much closer to the bone. What Bea and Vincent remembered about those times amounts to a vignette of past Cape Verdean life.

Vincent Pina's father had worked for Marion's wealthy, too.

He used to do a lot of these caretaking.... And that was seasonal work, he probably worked from May to September, that's all. And in the wintertime they cut wood. In the beginning they worked for, I think it was ten cents an hour.

To carry them through the rest of the year. Cape Verdeans relied on other resources, as Bea recalled:

In the meantime, the majority of Cape Verdean families also had a large vegetable garden.... I mean you had no lawns growing, you just had vegetables... squash, corn, peas, tomatoes, potatoes. Especially things that would stay over.... People did a lot of preserving.... So you survived that way.

Supplementing the garden crops were livestock, which Vincent's father kept, as did most Yankee families in Massachusetts's rural towns.

We had a lot of chickens. He used to have probably two dozen or better. Two dozen or better of hogs, unload them in the fall and just keep a couple for mating. And kill one for the family. Buy maybe half a beef. We used to have it hanging in the barn. After it's frozen, it would keep all winter.

Finally, staples and dried foods completed the store of food that got Cape Verdeans through the cold, slack season.

That's one thing about them old-timers, they knew winter was coming. You'd go in and buy a hundred-pound bag of flour, a hundred pounds of rice, a hundred pounds of beans, and some they'd [store] it upstairs in a room there. Crackers, everything. There were seven of us boys and we were never hungry.

Vincent and Beatrice Pina, Marion, Massachusetts. Lindy Gifford photograph

In the meantime, the majority of Cape Verdean families also had a large vegetable garden.... I mean you had no lawns growing, you just had vegetables...squash, corn, peas, tomatoes, potatoes. Especially things that would stay over.... People did a lot of preserving.... So you survived that way.

—Beatrice Pina

And how did life seem to Bea and Vincent, a life in which the little money earned was spent with great care, and there was no cash for extras? They answered together:

I think we lived better than a lot of people are living today. No money for movies. No cars. But there was a lot of card playing in the evenings. And a lot of neighbors visiting one another.... Sure, there was more togetherness.

Among Vincent, Bea, and Doris, only Doris remembered cranberry picking with any fondness. As a teenager in the 1940s, she worked each season. "And I used to be a good little picker. I used to be like a little snake. I was small, you know." Before the harvest began, there were certain preparations that an experienced picker always made. Doris and others would sew thick, cloth pads on their pants to protect their knees from the hard ground and rough vines they were constantly dragged across. Despite these precautions, a picker's knees were always raw and sore for the first week of harvest; it hurt even to place them against a mattress at night.

From day to day, there was always some excitement on the bogs to break up the drudgery of scooping.

There were fights, but they were still friends you know.... I would get mad myself, too, if you had to walk so far to get an empty box, to carry across—jump ditches, you know, and then somebody come and get it from you, you would get mad too.... But it was something that ended...they didn't stay enemies or anything like that.

Arguments broke out between pickers and the tally, the person responsible for keeping track of the number of boxes each worker picked. Pickers suspected that the tally never recorded their full number of boxes, while the tally assumed that the pickers were trying to inflate their pickings and earnings. Each was sometimes the case. To solve the problem, Doris recalled, "A lot of people used to get sticks of matches, every time [they filled a box] they put one in their pocket. You know, that's how they keep track of the boxes they picked."

Doris Gomes, Marion, Massachusetts.
Lindy Gifford photograph

Though Doris Gomes was a skilled scooper, and the bogs provided a little entertainment each day, the work's high point was its conclusion. "The best part was the end of the season, when you get your big, little check.... When the cranberry season was over you made three hundred dollars. Boy, that's a million!" For a hardworking teenager, these were good wages. Like most others, Doris turned her earnings over to her parents. "They'd give you a few dollars spending money and you'd be just as happy as can be with it." Doris's picking money, combined with that of her siblings and parents, bought winter food and clothing.

I really enjoyed it, though. Like I said, it could have been the crowd, you know. I used to have a lot of fun.... It's people I used to see, just during cranberry time, you know. Because they used to have truckloads coming from New Bedford. They'd go pick pickers over there. And, you know, you'd see them once a year.

—Doris Gomes

Even after Doris finished school, she returned to the bogs each fall. "After I got married, oh, for about fourteen, fifteen years, that's all I did was pick cranberries. I never worked [full-time] after that. My kids, you know. For that six weeks, I found a babysitter." The harvest also cemented friendships among Cape Verdeans who converged on the region yearly from distant places.

> I really enjoyed it, though. Like I said, it could have been the crowd, you know. I used to have a lot of fun.... It's people I used to see, just during cranberry time, you know. Because they used to have truckloads coming from New Bedford. They'd go pick pickers over there. And, you know, you'd see them once a year.

They came not only from New Bedford, but from Worcester; a neighborhood in Cambridge; the Fox Point section of Providence; Norwalk, Connecticut; and even Rochester, New York.

No one was excluded from the cranberry harvest, not even the very old. It was not uncommon to find men and women well into their seventies scooping daily. Lincoln Thatcher, an affable grower from North Harwich, worked alongside an elderly

The harvest.
Ocean Spray Cranberries, Inc.

Cape Verdean woman on the first day he ever scooped. At picking's end, Linc—blistered and sore—had scooped sixteen boxes. His aged neighbor had picked eighty-seven boxes of cranberries, a skill she had developed over a lifetime. Loyalty also motivated older Cape Verdeans to return to the bogs each fall, as was the case with Bea Pina's grandmother.

> You know, everyone would say, "Now you stay home, Minnie, because you know, you're too old to go picking." But she had to go, because she had to pick for Makepeace. Nobody was going to keep her home. Makepeace was God to her.... Even to the day she was dying, she always praised the Makepeaces, because that way she was able to get social security. He did something, I mean he paid social security on her.... So she was collecting social security. It wasn't much, but she had her little check coming in every month. She praised Makepeace, because if it wasn't for him, she'd never be getting that little bit of money.

Despite the punishing difficulty of the work, the harvests were, as Edward Garside wrote, "gala days." In the growers' shanties and in Cape Verdean homes, people caught up on what had happened in their lives over the months since the last harvest. Tables were pushed aside, and there were kitchen dances and all-night parties. And the harvest was a time for christening, welcoming babies into the world. For it was then, as at no other time in the year, that there was money about.

Tony Jesus, Onset, Massachusetts.
Lindy Gifford photograph

Chapter 7

Tony Jesus

The village of Onset in Wareham is a place with two identities. A fashionable resort on the warm waters of Buzzards Bay in the late 1800s, it is still a summer home for middle-class whites who return to the small Victorian cottages year after year. Onset is Wareham's largest Cape Verdean neighborhood as well, a settlement far back from the shore, where streets are narrow and lined with closely built houses—summer cottages made over for year-round living and a few bungalows. On Thirteenth Street was Tony Jesus's home. If you arrived there early enough, you would find Tony out for a walk in the neighborhood, managing remarkably well for a man in his nineties with an artificial leg. Come later in the day, and Tony would be inside with neighbors and family. Constant visiting between homes is a hallmark of Cape Verdean life and makes for households that are friendly and open.

On a day early in spring in his small living room, Tony Jesus talked about his life, amid calls from his sons and several friends stopping by to say hello. In many ways, Tony Jesus's life was the life of most first-generation Cape Verdeans who arrived in America in the early years of the twentieth century. But in places, the lives of the one and the many diverged, and Tony Jesus, clever and eager to learn, did things others would not do.

It was in 1902 that Tony Jesus, then a young boy, boarded an ancient schooner in Brava and sailed across the Atlantic Ocean to New Bedford, Massachusetts.

My mother was here then. My mother come three years before, then she send for me and my brother, which is the one next to me, was four of us. One took up a cook and me, I took up everything I can find, and one next to me took up a mechanic, and the third one—his picture up there—he took up a law. He studied, he went into law business for forty-two years before he died. In the family, I'm the only one living.... And of course, education was small, 'cause I want to travel. That was my mistake, which I come to know, that education amounts to something. Man or woman got an education can find a good job. Man or woman without an education have to scrub floor for your living. And which is right up to today still. So I travel 'round. I went as far as the third grade, I got out of school, that is I run away from the school, and I went to look for a job and I found a job. In a factory, a cotton factory, which that factory still in New Bedford, they call Wamsutta Mill, up North End.

The whale went down, and the officer said, "That's the worst one I see in all the year that I been on whaling business." He said, "Everybody watch out." And by God, before we know, he come right underneath the boat, he took the boat up in the air, smash it all to pieces, everybody overboard.

—Tony Jesus

Tony soon got his chance to travel, for at age fourteen he signed on the bark *Sunbeam,* a whaler out of New Bedford, as a cabin boy. But Tony did not find the life of a cabin boy to his liking. He spent his time setting tables and performing galley chores, when what he wanted to do was join the men in a whaleboat and see how the whale was taken. Despite much pleading on Tony's part, the captain would not allow it, knowing that the fourteen-foot oars of a whaleboat were too much for a boy to handle.

The voyage continued. Filled with whale oil after four months, *Sunbeam* put in at Montevideo, Uruguay, to discharge the oil and take on fresh stores. The whaling began again and the next port was Martinique, where Tony began his master plan to get aboard the whaleboat. While on the island, Tony hired a boy to take his place in the cabin and brought him on board, unbeknownst to the rest of the crew. The unfamiliar face on board was soon discovered, and Tony was made to confess his doings to the captain. Perhaps impressed by his ingenuity and determination, and uncertain how else to use the newest crew member, the captain let the plan go forward. Tony Jesus became a deckhand. He now set and trimmed sails and, high up on the masthead, scouted for whales.

Your spyglasses, you looking for whale. When you see a whale, you holler. And then captain, if he's in the cabin, he come out, he ask you, "What you see?" "I see a whale." "What's he doing?" You tell him what the whale is doing, if he's rolling, if he's traveling along or what he's doing, you tell him. Then you know what to do. And then of course he'll get up there and watch everything's all right. Then he'll holler, "Get the boat ready!" Boat's all ready, you know, all ready to go. Only the mens on the boat. There's five men to every boat. The boat steerer and three row men and a harpoon man. So everybody gets on the boat. "All right, load away," they yell, "get in the water," and they go after the whale.

To go after the whale was still Tony's burning desire and now, as a young deckhand, within the realm of possibility.

And so I asked one of the men that was great friends with my father, I asked him then. He said, "When we get ready to go," he said, "you jump in my boat." So I did. You know, just for hard luck, first time we get on the whale and we got smash-up. And the harpoon man put the harpoon on the whale, the whale went down, and the officer said, "That's the worst one I see in all the year that I been on whaling business." He said, "Everybody watch out." And by God, before we know, he come right underneath the boat, he took the boat up in the air, smash it all to pieces, everybody overboard. Me, I know how to swim, I didn't care. To show you, sometimes young boys we crazy. I'm swimming around. You can't see no land, just sky and water. And the ship is about two miles away from here. Right away they send what they call a spare boat, aboard the ship, to pick up the mens. You know, I look around, I see the officer is dog-swimming. I said, "Hey, that man don't know how to swim." I reach over one of the boards that come from the boat smash-up, I took it, I put right before him, he grab it. He didn't forget.

The whaleboat officer did not forget. Back on *Sunbeam* in his cabin, he rewarded Tony for saving his life with a quarter pint of rum, poured from a jug. It was Tony's first drink of alcohol and he got

Smith-Fuller-Hammond Company teams, Main Street, Wareham, Massachusetts.
Fuller-Hammond Co.

drunk, sleeping it off for the rest of the day in his bunk. Next morning, he joined the crew in cutting the blubber and trying it out for oil.

Back in New Bedford after two years at sea whaling, Tony Jesus had still not had his fill of traveling. He shipped out for a year or so on a schooner that traveled the New England coast, carrying sand and other bulk cargo from New York to Boston. Then it was on to a coal barge for several years.

And I traveled around so I got so I wanted to get married. Well, I didn't have but one parent. So asked my mother I was going to get married. She died in 1972. And I liked the farm work. So I married and, just as I said, the first year, in 1919, I came here. I drove team for the company.

Many Cape Verdean men of Tony's generation brought to the comparatively tame world of the bog memories of hunting whales or seeing strange sights from the deck of a schooner traveling the Atlantic coast. The company Tony Jesus came to was Smith-Fuller-Hammond Company, cranberry growers who established large bogs in Plymouth County early in the century and eventually evolved into two separate companies known as Smith-Hammond and Fuller-Hammond. Tony approached Irving Hammond, part-owner and manager of the bogs, for a job and immediately displayed considerable intelligence.

And so I come look for a job. He say, "Sure." But I said, "The job I want, I don't want picking, scooping, I don't want that." Well, I figured—I'll show you the picture over there of scoopers—you put your knees down on the ground for six, seven hours, just pushing, pushing along. Boy, that's hard work. I said, "If I can get away from that...." I know how to do it, can teach you how to do it, but I don't want to do it myself. Okay. So I said, "The job I want probably got a man already." He said, "What kind of job you want?" I said, "Teamster." "All right," he said, "I'm looking for a man that drives horses." So I got the job.

Ditching.
Ocean Spray Cranberries, Inc.

If there's going to be a frost tonight, I've got to work all night.... You flood the bogs, you know, nights, to protect them from freezing.... You fool around all night till sunrise in the morning.... You got to go back set the men to work..... Oh, sometime for two nights and two days you don't take shoes off your feet.

—Tony Jesus

As a teamster, Tony stayed off his knees and avoided the punishing labor of scooping. His job was to collect the boxes of cranberries at the bog and deliver them to the screenhouse for packing. When picking season ended in late October, Tony and many others returned to New Bedford and work in the cotton mills. The following year when Tony Jesus returned to the bogs, it was for good.

This is 1920 I start in year-round man. This is on the Smith-Hammond side, but we worked both sides, Fuller-Hammond and Smith-Hammond.... And then he got so he thought I was pretty good on handling the men, so he had me taking the crew 'round. Because the company had a bog about everywhere. Got a bog down East Falmouth, got a bog in Plymouth, near Plymouth Town, got a bog in North Easton, that's the other side of Taunton. A hundred twenty-seven acre up there, which he still got it. That's on the Fuller-Hammond side.

Tony Jesus was soon a foreman for the Smith-Hammond Company. His special responsibility was Carver Bog in South Carver, and he and his family lived in a company house close by, where they had a garden and kept a cow. It must have been an isolated life, for in this area, far removed from town and people, the Jesuses' only neighbors were great expanses of pine woods and acres of cranberry bogs.

The responsibilities of being a bog foreman were many and never so important as in spring and fall when the fate of the crop rested in Tony's hands. Frosts are common in both seasons, killing the blossoms early in the year and damaging the berries around picking season.

If there's going to be a frost tonight, I've got to work all night. And then those days, we didn't have no sprinkler like we have today. You have to walk. I had twenty-seven flume, I had to walk to every one of them. You gradually flow. You flood the bogs, you know, nights, to protect them from freezing.... You stay all night. You fool around all night till sunrise in the morning.... You got to go back set the men to work. And then you got to start the pump and pump that water back.... In case you have to use it again, you got water.... Oh, sometime for two nights and two days you don't take shoes off your feet.

When cranberries were ripe and the harvest was on, Tony Jesus's duties were only a little lighter. On weekdays there might be fifty scoopers on the bog, and Tony would handle a gang of twenty-five men. He would see that scoopers filled their boxes with berries, not vines covered with a thin layer of fruit, and that squabbling between scoopers over empty boxes stayed friendly. If Mr. Hammond decided to pick by hand in order to avoid damage to the berries that scoops sometimes caused, the number of pickers doubled, and the bog became a circus.

I've seen over one hundred people between mans and womans and kids on Saturday and Sunday. No school, the mothers bring their kids. I've seen more than a hundred head right before me, picking by hand. They have measures, six-quart measure. They have measure right side of them, they have measure in front of them, they pick by hand. When this get full, put there, get the empty, put it here, keep going. And the kid take it and take 'em to the girl that mark 'em down.

For the most part, though, Tony's pickers were men who pushed the scoop along for the month or so that harvest lasted. Their homes for the season were the buildings erected and owned by the company. "Little shanty they call them," said Tony, "what we call a summer house, you know." The shanty that Tony remembered best, peopled with faces out of his past, is at the Fuller-Hammond bog in North Easton. There, down behind the screenhouse and at the edge of the bog, stands "Fish Market."

Why they name him Fish Market, because the grocery man come there, fish man, meat man, that's the first place they stop.... There is going to carry about eight ten men. So name it Fish Market. So they live right there, one big kitchen, upstairs, all bunks, whole side. And 'course, real cold night, they used to keep one man up, say two hours, he stay from ten to twelve, put the wood in the stove. And then he'd call the other guy, the other guy come down, and he'd go to bed.... They get along fine. Only they don't pay no rent and the wood they don't have to buy.

With the last of the late Howes berries scooped, screened, and packed, most harvesters returned home to other jobs or a long, quiet winter. But a few men always stayed on, working at the jobs that could be done before the cold came and the bogs were flooded. On many autumn days, Tony and his crew could be found ditching, digging the sand and muck out of the ditches, and paring the vines along the edge so that the winter flood would travel around easily.

After picking season we have a lot of ditches that filled in, you know. So we put the men there with the shovel. And we have a plank right side of the ditch, and you can't wheel on just the vine, you had to wheel on the plank. And mens on the ditch they cut the edge and put 'em in the wheelbarrow, and a man wheel 'em ashore. If we get all done then we go

Sanding.
Ocean Spray Cranberries, Inc.

ahead and do some sanding. 'Course you don't put on heavy coat, you put on light coat. Like this they sand, say, so many acres. Next spring you sand so many acres. Next fall, same thing until you catch up.

Since manual labor was how work got done most of the years Tony cared for bogs, there was no catching up. Spring brought more ditching and sanding, and in summer there were insects to spray and hand weeding to be done. In either season, new bog could be built or old bog rebuilt, a big job in either case.

Oh yes, I built many acres for the company. Oh yeah, and rebuilt good many acres for the company. Sometimes they get old and the vine begin to die out, 'course you have to take the top off and rebuild. Like you say, suit of clothes. You got a suit of clothes ain't much good, shaggy, why, take 'em to the shop, they fix it for you. Same thing as the vine.

So the cycle of work on the bogs went, year in and year out. Tony Jesus was well liked by the men who worked under him, by the families who came from distant towns to pick cranberries, and by his boss, Irving Hammond.

He was a good man, but he liked to swear. Oh, boy. Of course most of the people, old Yankee, like that anyhow.... Because Mr. Hammond, God bless his soul, he was a man that damn the man. He'd damn you just as soon as look at you. He can't say that without "God damn" first. Although it's kind of hard word to say. Well, that was his way. "God damn the nation!" One day I said, "Mr. Hammond, what do you mean, 'God damn the nation,' what nation you talking about?" He laughed then, when I told him that. Sometime he take my joke, he laughed, you know. He was an old man.

But he used me good. I was number one in the company, oh yes, number one in the company. Because, well, I used to do all I can. Although, let me tell you something, a lot of people is that way. Don't make any difference.... I remember, a lot of times I carry mens, I got ten, twelve men, I go from one bog

Mr. Hammond used to say, "God damned drunkard! Fire him! Fire him!" That's always the word, "fire." I don't want to fire the man. He's all right, after he get sober, he's all right. So that's why I get along.

—Tony Jesus

to another. And they got their pint before we started. By the time we get to where we're going to work, one of the guy is drunk, he can't wheel sand. That don't worry me. I say, "Wait a minute. Put him over there. Go sit down over there, I'll take the wheelbarrow." I do the job for him. I don't get any more for that. I said, "I'll watch for Mr. Hammond, if he come, why, you get up and stand over there." But I watch him, because Mr. Hammond used to say, "God damned drunkard! Fire him! Fire him!" That's always the word, "fire." I don't want to fire the man. He's all right, after he get sober, he's all right. So that's why I get along.

You know, every once in a while, just as I said a little while ago, once in a while somebody say to my boys, "Tell your father we say hello. He's a wonderful man. By God, he's still around? You can't kill that fellow." So you see, it pays to be good in this world, because when you die, you dead and gone and you ain't going to take nothing with you. Yes sir, that's the way I feel. I want to be good to everybody. Which I have a lot of good friends. I know a good many people, all kind of nation. I traveled Finn, I traveled Italian, which I got a lot of Italian friends right up here in Tremont. And Polish.

Tony Jesus retired as foreman for the Fuller-Hammond Company at age ninety and after sixty years on the job. He did not retire because he wanted to, but because he had to. In 1980, an infection in one of his feet resisted medical treatment and soon became a leg infection. Despite the care of Boston doctors, the infection could not be purged, and Tony's leg was amputated. So at ninety, Tony Jesus weathered a major operation and learned to walk all over again. The infection, Tony attributed to the pesticides and herbicides used on the bog. "You know, people won't think so, but you see we use a lot of poison in the bog which is no good for man's foot, or your body, might as well say. But I don't know, we have to take it and like it."

Just as I said, if it wasn't because I lost a leg, maybe.... I was going to try to go just as far as I can. Because when you get a certain age you get fired anyhow. But on bog they don't fire you. Work all you want to. Work till you drop.

Bogs at the Harju homestead, South Carver, Massachusetts.
Lindy Gifford photograph

Chapter 8

From Finland

It was from Finland that the only other ethnic group to settle permanently in southeastern Massachusetts and work the cranberry bogs came. And as in the Cape Verdes, it was sailors who first saw America and returned to Finland in the 1850s to reveal that the United States had what Finland did not: industry that needed workers and land that could be purchased cheaply. For the Finns, the lure of America stemmed from serious problems in their own economy and social system. During the nineteenth century, Finland experienced great population growth combined with dwindling opportunities for its young. In rural Finland, farmsteads had always been left to a single male heir. With a growing population, an entire class of younger sons with no property and no stake in Finland's future was established. Landless agricultural workers increased more rapidly than landowners and were dissatisfied with their lot in life and prospects for the future.

It was these men and women, Finland's dispossessed, who began to appear in Massachusetts in the 1870s. Fully 90 percent of Finnish immigrants to America came from rural districts, largely the northern coastal provinces of Vaasa, Turv-Pori, and Oulu, where word of America's opportunities came early and access to ports was easiest. For Finns the voyage to America held few of the dangers it did for Cape Verdeans. They departed Europe in the relative safety of a steamship.

In Finland, the ocean journey began with a trip to Stockholm, or later, to Hanko, Finland. From these ports, steamers departed for Copenhagen, Lubeck, or Hull, where the final leg of the voyage would begin. Many Finns traveled on prepaid tickets sent to them by family or friends already in America. The trip must have been less frightening for these travelers, knowing that there would be familiar faces at the other end to help ease them into life in another land. But no one escaped the unpleasantness of steerage travel. Living, eating, and sleeping among strangers in the ship's hold, there was no way to avoid the odor, sickness, and depression. The blond Finns felt a superiority to the darker eastern Europeans, and this made itself plain during the voyages when the Finnish withdrew to themselves. This trait of keeping apart from other cultures was one that native New Englanders soon noticed when the Finnish settled in their towns.

Quebec, Boston, and New York were the common ports at which Finns disembarked to continue their travels by train to final destinations. For many arriving in Boston, the train ride was not long. In the years between 1893 and 1905, more than 25,000 Finns came to Massachusetts. Only the state of Michigan attracted more pioneer Finns, and Minnesota and New York claimed nearly as many. With the outbreak of World War I, large-scale Finnish immigration to America came to an end.

The Cape Verdean and Finnish experiences in southeastern Massachusetts diverged quickly. Certain natural advantages and their particular cultural heritage helped Finns to establish themselves more quickly than their black neighbors. The Finns were white and western European, with a rural, agricultural background. The immediate social division based on color that existed between Yankees and Cape Verdeans did not occur for them. Finnish newcomers endured the epithet "squarehead" from natives, but racism never went beyond name calling. As well, the Finns came out of the same general cultural tradition as the Yankees they settled among, whose ancestors had left an agrarian life in England nearly three hundred years earlier. The particulars of Yankee and Finnish culture differed, but not the basic orientation. For Cape Verdeans, whose life ways were more West African than European, the cultural disparity between themselves and Yankees was pronounced.

Things in the recent past of Finnish immigrants also worked to their advantage when settling in New England. On arrival, most Finns were forced to accept jobs that were readily available—those in industry. They found work in marble quarries, foundries, furniture plants, textile and wire mills. The Finnish presence was and still is especially strong in the cities and towns of central Massachusetts, centered around Fitchburg. But when they had saved a sum of money, earned over several years, the Finns quickly returned to their traditional occupation: farming. In a time when Yankees throughout New England were abandoning farms for city life and the higher wages that industry offered, Finns reversed the trend. They bought—for little money—Yankee farmsteads and began to raise poultry, crops, and fruit. Many who came to the cranberry-growing towns had worked there in previous seasons on the harvest, learning a little about cranberry cultivation. They returned for good, bought an old house, a couple of acres of bog, and tried their luck. Both luck and a formidable work ethic often brought the Finns success. No strangers to deprivation at home, they overcame whatever difficulties American life brought them. *Cranberry Red* author Edward Garside observed of the Finns, "But they were the hardest themselves, they were as hard as nails. Oh, yes."

The Finns who emigrated to America were a generation dispossessed of their rightful heritage, land. They knew the value of property and of ownership and bought land as quickly as possible after arriving. Farming was in their blood, it was all that many Finnish families had done for generations, so it was natural that they should return to farming in New England. Used to having less, a Finnish family could make do on a tiny farmstead with resources that most Yankees considered a little too thin.

In the cranberry-growing region, Cape Verdeans rarely bought or built their own cranberry bog. Although many purchased or built small houses, reminiscent of bog shanties, few invested in bog property and did the kind of work for themselves that they did for others. While Finns soon established themselves as bog owners, Cape Verdeans continued to labor as bog workers. Once again, cultural conditioning may have influenced this outcome. The Finnish came out of a tradition of owning land; Cape Verdeans came out of a tradition of working it. In the islands most had worked

Cranberry box shipping label.
Private collection

So [my father] came out here. And he came out once just for the fall to pick.... Then he came back...and he bought a house and farm in 1910. It had a little bit of bog on it.... It was never a very big bog, five acres, five and a half acres.... They had cows and grew their own stuff for what they had to eat.... Cranberries were the cash crop to help out a little bit. And we all started from there, my brothers and I.

—Wilho Harju

as laborers, or at best, as tenant farmers, for large landowners. Life had been that way for Cape Verdeans since the Portuguese started sugar plantations in the sixteenth century. The custom of owning land was not instituted and the compunction to do so, once in America, was not strong.

Other factors worked to keep Cape Verdeans from establishing themselves in the growing region as the Finns did. Cape Verdeans always maintained a certain ambivalence toward life in America; many thought their time here would be temporary. Often, their goal was to return to the islands with money and live well. With this plan in mind, many Cape Verdeans may not have wanted to encumber themselves with the responsibility of a bog, or to set down their roots too firmly here.

And there was the matter of literacy. Of Finnish immigrants to America age fourteen and above, only 1.3 percent were illiterate. Among Portuguese immigrants, which included those from Portugal, the Azores, and the Cape Verde Islands, the illiteracy rate was 68.2 percent. Literacy in one language was very useful when trying to learn another. A more rapid ability to communicate and read in English must have helped Finns enormously in becoming citizens and property owners. Cape Verdeans acquired these tools for Americanization much more slowly.

The story of one Finnish family's journey to New England and experiences here parallel with amazing exactness the general scenario of Finnish emigration to America. Wilho and Lillian Harju live among hundreds of acres of bogs in South Carver, not far from the small Cape Cod house that is the Harju homestead. Wilho's parents bought it almost a century ago, their first real home in America.

They were born in Finland. They came to Worcester in 1903, I think my father did, or 1904, in that time. And then they got married in 1905, and they lived there for seven years. And on account of health reasons, they told my father that he'd better get out in the country, find some occupation that.... He was working in a wire mill, steel and wire mill, in Worcester. He contracted some kind of disease. I mean, his health just went bad on him. He was losing a lot of weight and they said he'd better get out and get some fresh air. I guess they had a lot of acid and stuff in the wire and stuff that's bad for your health, your lungs.... So he came out here. And he came out once just for the fall to pick and then he went back, spent the winter there, worked there. Then he came back through fall and spring, and he stayed here and he bought a house and farm in 1910. It had a little bit of bog on it, probably an acre and a half. And that was some of the oldest bog in Carver, according to some of the records.... And he expanded on that. It was never a very big bog, five acres, five and a half acres. And that's what they made their living off of, besides kind of living off the land. They had cows and grew their own stuff for what they had to eat.... Cranberries were the cash crop to help out a little bit. And we all started from there, my brothers and I.

I used to take floats in the fall.... Whatever dropped on the bog...they'd flow them up and pick up the berries in...float scoops.... But it was fairly good business and we used to split the profits with an owner.... We had all these little trades in between plus growing them, so we got along all right.

—Wilho Harju

When Wilho Harju's father arrived in Carver, he found that he was not alone. Other Finns had preceded him there.

There was like an older generation. A few of the people that were older than my father, maybe fifteen, ten years older, that had been here almost ten, fifteen years. Some of the real old settlers had come here in the 1880s, late 1800s, to the Carver area. And a lot of them were contracting [building bogs under contract] and some of them were foremen for some of the growers.

A lot of them came over in different ways, some came from the cities, and some were seamen that came off the ships that came out of Boston or somewhere. I mean, heard of the opportunities out here, so they came, which they thought were opportunities. Cranberries grew in Finland...they weren't grown there as a commercial crop, but they were known what they were. [It is the lingonberry, or mountain cranberry, that grows in Finland and through much of Northern Europe.] They were used as a food, so the Finnish people all knew what cranberries were. They weren't something strange to them. Of course the cultivating of them I guess probably created a little interest for them. You know, here they're cultivating them and back home we used to just pick them in the wild and so it would probably be a kind of interesting thing to get into.

Since that time, the Finns have had a reputation among Yankees and Cape Verdeans as "good growers," the industry's superlative for exceptional cranberrymen.

Like many Finish immigrants, the elder Mr. Harju and his wife-to-be had lived in the country. In fact, they grew up in adjoining towns, though they did not meet until both were settled in Worcester, Massachusetts. Unknown to each other in Finland, they soon found they knew many people in common. What were the specific conditions that caused Mr. Harju, his future wife, and so many others to chance life in a country an entire ocean and more than 4,000 miles away?

To try and improve their lot, that's all. I mean, in most of those countries the oldest son inherited the property. I think it's true in more places than in Finland. I think it's true in all of Europe, at that time, anyway, that the oldest son inherited the property, with the understanding that he takes care of his parents when they get old. And so if you were some of the younger ones, why you had to kind of shift for yourself and start looking.... Well, you got to find a place. Either you could stay on there and work, kind of for nothing and be a, well, practically a slave to somebody the rest of your life, or then you could go

Taking floaters. Middleborough Public Library

and try to better yourself. Over there the opportunities weren't that great, so I suppose to everybody that came from Europe, this was the golden country, where you picked the gold off the streets, so that's where they came, you know. I know my father said you'd never catch him going back there to live. He said he wouldn't mind going back for a visit—he never did—but he always said he'd never go back there to live. This country was by far the best of anything he'd ever heard of or been in.

Mr. Harju found what he was looking for: a place where he was his own master, could work his own land, and better himself.

He lived to be eighty-five, by the way, so he had the bog right up until the day he died. We used to take care of it for him. He never expanded above his original acres there and he made a good living off of it, so he didn't want that much in life anyway. He was warm and had clothes, he was comfortable, very comfortable. And that's all he was looking for, he didn't want to be a millionaire or anything else.... He was comfortable and he was to the age when he was glad to see us branch out a little and get out on our own. [The Harju boys all bought bog of their own.] I know he thought quite a lot of that, when we got started on our own, he thought we were doing pretty good.

Wilho Harju's start in the cranberry business was as modest as his father's. Out of the service in 1946, he bought a small bog in the waning days of the wartime cranberry boom and saw his profits dwindle to almost nothing within four years. It would be thirteen years before Wilho, with a wife and family, could afford to become a full-time cranberry grower.

Wilho and Lillian Harju. Lindy Gifford photograph

I'd be working all day weeding out there and then I'd come home and make supper and then my husband and I would go and we'd work until the mosquitoes ate us all up. Then it was time for us to come home because we couldn't see another weed, so dark you know. We would do this day after day until summer was over and it was harvesting time and—yeah, we did a lot of work there.

—Lillian Harju

I was a carpenter for years and I set vines for years. I used to set vines by the old hand method. That was before the machine method. And I had a group of women, as many as fourteen women—Cape Verdean women—working for me. And we used to set an acre or an acre and a half a day, with that crew.... I set sixty-four acres of new bog in 1947.... And I used to take floats in the fall. I used to be one of these that picked with a hand scoop. Whatever dropped on the bog in them days, which was considerable in some places where the vine was real bad, they'd flow them up and pick up the berries in the same way that we water-pick today. Only we did it manually more, we had these float scoops that we used to use, wash them out in those, instead. But it was fairly good business and we used to split the profits with an owner and we did the work and pick 'em up and get rid of them—sold them and we split with the owner. It was kind of an extra income for us, so we did that. We had all these little trades in between plus growing them, so we got along all right.

In the many years of low cranberry profits and "all these little trades in between," Wilho Harju never once considered leaving the bogs.

No, we were kind of stubborn and we'd been in the business all our lives, and it was kind of the only thing you really liked. And if you did carpentry work, you liked it in a way but you didn't like it as much as growing cranberries. You always figure that well...next year's going to be better. That's what kept us going.

Nineteen forty-nine was the golden year for Wilho Harju, a year of long-awaited independence.

Nineteen forty-nine, we bought this property up in Plympton, the old Paul Thompson Bog, and that summer I quit and I never went back to work for anybody, I worked just for myself. I've never been sorry, either.

Perhaps more than many, the Harju Bogs have always been a family operation, a characteristic of the Finnish growing families. For years, this included Wilho's wife Lillian, who, though not Finnish, was blessed with the Finnish capacity for work.

They used to tell me I was crazy to do this because they wouldn't do it. But there's a lot of women who have worked, I'm not saying they haven't. But they say they married a man not for that type of work. It's really hard work.... We'd come home—well, I'd be home for supper you know, I'd be working all day weeding out there and then I'd come home and make supper and then my husband and I would go and we'd work until the mosquitoes ate us all up. Then it was time for us to come home because we couldn't see another weed, so dark you know. We would do this day after day until summer was over and it was harvesting time and—yeah, we did a lot of work there.

Harvesting time found Lillian Harju back on the bogs with an advantage of few working mothers: her children nearby.

Well, when I scooped I'd scoop about fifty-four bushel of cranberries for my husband and myself. And I would take the boxes from shore early in the morning and put 'em up on the section and put the children out to play. If I had a little one, he'd be sitting in a box and I'd be harvesting with my sister-in-law.... Children never too far from you. A lot of people used to say my goodness, it's an awful place for your children, but I never found it was hard for me to have the children around the bog, because if you took them to the beach they'd be playing in sand. Well, if you took them to the bog they had the sand right there and then they had the little ditch and they used to play in the water.... No, I think I was kind of rich that way 'cause I knew where my children were.

Unrecorded by the U.S. Immigration Service, unnoticed by townspeople, the Finnish were not particularly conspicuous in the cranberry-growing region. They blended into the cultural landscape, outwardly different from their Yankee neighbors only in the name on the mailbox.

Ocean Spray ad.
Private collection

Chapter 9

The Raw and the Cooked

The Cranberry Cooperatives

In the cranberry industry's first fifty years, from 1850 to 1900, the cranberry grower was an extremely progressive agriculturist. He had succeeded at the risky task of bringing a wild plant into the realm of cultivation. Through observation and innovation, the grower learned to meet the cranberry plant's physiological needs and make it more fruitful in captivity than it had been in the wild. To match the new fruit, men invented new tools and machinery for growing and picking the crop. Within the tradition-bound world of agriculture, these were great accomplishments for a scant half-century. In 1907 the cranberry growers of Massachusetts took the last major progressive step they would take for another fifty years: they formed a cooperative.

As the final years of the nineteenth century played out, problems became apparent with the established means by which the growers sold their berries. By tradition, they dealt with some commission merchants through the mails and with others through their agents who visited the bogs. Whether the going price for cranberries was quoted on a postcard or by an agent, growers shipped their crop, hoping the commission merchant would do well by them, and waited for the returns. What sometimes happened was recounted by the late Larry Cole, a grower whose family had long been associated with cranberries and the barrel and box business in North Carver.

Let me tell you a story, and this is why we got the farmer's co-op in the beginning. I'll tell you two stories, and they're true ones. First my great-grandfather shipped some berries to Baltimore, Maryland, before 1900 and he waited for a check that never came. Just kept waiting and it didn't come.... Those were the horse and buggy days, you didn't know who you were doing business with, except by letter; there wasn't the communications we have today. Maybe that man down there had tough luck and maybe he didn't, but he knew very well that Ben Robbins wouldn't find his way 500 miles to see why he didn't get paid for fifty barrels of cranberries, more or less. I can't tell you how many; he wasn't a big grower. So he got them for nothing. Well, my father would chuckle as he told me how his grandmother used to scold him [Larry's great-grandfather] every fall. "Don't you ship any more berries to that fella. You ought to have known better." Well, how was he to know?

Well, there was a fellow up in Rocky Meadow Cranberry Company, and that was the bog that George Olsson bought, well, later on. His name was Ben Shaw, and he took care of the bog. He shipped some berries to the commission house in New York. My father's told me this story many times, and it's true. He got a check back, he was fortunate in that respect, but the check was much too small. He knew he hadn't been treated fairly. So there were trains—the trains went through Carver in 1890 something, early 1890s. He took the train right away. He went to New York, and he dressed up and give the impression that he had a store about twenty miles outside the city of New York. And he went into the produce market and he wanted to buy things to sell retail in the store. I suppose a sack of potatoes and onions—and he come across cranberries, yes, he wanted to buy some of them. And it wasn't long before he was bargaining to buy his own cranberries, but the commission man didn't know it; the only way they'd ever met was by letter. And the price for those berries, he couldn't believe it, it was so high. So he was going to negotiate, get the price down, but the fella wouldn't lower the price. He said it's a good market.... "All right," he said, "I want 'em." And as he went to pay for them, he reached in his pocket and he showed him this letter and the check that he'd got the week before, and he said, "You tell me one thing. How come these berries can be worth so much this week and so little last week?" He said that fella's face was as red as the berries.

Many cranberry growers had cordial relations and even friendships of long standing with commission merchants in distant cities. But there were others who took a chance and shipped berries to merchants they hardly knew and were preyed upon in return. George Briggs, a Plymouth grower speaking in 1919, recalled for younger cranberrymen the other great marketing problem that existed before the cooperative:

I can remember well the result which invariably followed a tour of the Cape by an agent of a Boston house that was well known in those days. In this case there was no attempt to deceive the grower about the probable price. The price was easily gained if shipment was immediate, but nearly all growers were using the same haphazard method, and in a day or two, Boston would receive more berries than its dealers could sell in several weeks and it fell from the highest to the lowest market in the country. So it was to a considerable extent in New York, but the outlet was better there and the market did not fill up quite so rapidly. There was no systematic distribution in those days. We filled to satiety the various markets in turn and, in consequence, every market had its ups and downs.

It became increasingly apparent that in order to avoid unscrupulous merchants and the pitfalls of haphazard marketing, growers needed to control cranberry distribution themselves; they needed to cooperate. Though new, agricultural cooperatives were not unknown in New England. In the years following the Civil War, small dairy cooperatives were established in New York, northern New England, and Canada. But no successful cooperative had yet evolved in New England for the marketing of vegetables or fruits. As early as 1895, cranberry growers in New Jersey and Massachusetts had dabbled with cooperative selling methods. Large producers in New Jersey and Massachusetts formed the Growers' Cranberry Company and hired a salesman to place their crops. Other Massachusetts growers did the same, incorporating the Cape Cod Cranberry Sales Company. Although these growers did join together to sell the fruit, neither their berries nor their profits were pooled together. Each grower's crop was still sold under his own label, and he received the returns from only those sales. Cranberrymen remained leery of wholly cooperative methods; each grower still wanted to retain his independence, his label, and his profits. Representing only a few growers in each state, neither company was particularly successful. In order to control the flow of berries to markets and maintain a good price for them, it was necessary for many cranberry growers to sell their crops together.

It took several years of enormous cranberry crops, during which the cost of production exceeded the selling price, before growers finally conceded the need for a real cranberry coopera-

He reached in his pocket and he showed him this letter and the check that he'd got the week before, and he said, "You tell me one thing. How come these berries can be worth so much this week and so little last week?" He said that fella's face was as red as the berries.

—Larry Cole

Cranberry barrel labels.
Private collection

tive. And then, it was not the producers in Massachusetts who acted first, but those in Wisconsin. From that time forward, the cranberry growers of Wisconsin have always proved themselves more progressive and open to change than their Massachusetts or New Jersey counterparts. In 1906 the northern growers formed the Wisconsin Cranberry Sales Company, a cooperative open to all producers willing to pool their crop with that of others. Ninety percent of the growers in the state joined. To run the co-op, the growers chose Arthur Chaney, a wholesale grocer from Iowa, who had handled Wisconsin cranberries in the past and was regarded as capable and scrupulously fair. Within the co-op, all cranberries were pooled by variety and carefully graded. Each grower received the season's average price for the variety of berries he had contributed. Thanks to cooperative marketing, Wisconsin cranberry growers made a profit in a year when a bumper crop had to be sold.

RULES FOR BRANDING

General Requirements. All cranberries which are branded must be dry, sound, free from frosted or wormy berries and — with the exception of "Variety" Brand — free from berries of a **green** color; unless specifically mentioned they should not contain any **all** white berries; they must be solidly and cleanly packed and reasonably uniform in size. Pie Berries must be removed by the use of a 13/32-inch grader.

Special Requirements of the Several Brands

EARLY BLACKS

Chanticleer — Not Eatmor
Blacks; at least 90% colored; count from 125 to 150; fit for 15 days' travel.

Skipper — "Eatmor"
Early Blacks; averaging 75% colored, with not over 10% of all white berries; counting not over 130 to the cup; fit for 15 days' travel.

Mayflower-Lion — "Eatmor"
Blacks; at least 90% colored; count not over 125; fit for 15 days' travel.

Minot's Light — "Eatmor"
Fully-colored Blacks; count not over 125; fit for 7 days' travel.

Capitol — "Eatmor"
Blacks of uniform red color; count not over 140; fit for 15 days' travel.

Yale — "Eatmor"
Blacks; 90% colored; graded over a ½-inch screen; count not over 95; fit for 10 days' travel.

Harvard — "Eatmor"
Blacks of uniform dark color; graded over a ½-inch screen; count not over 95; fit for 7 days' travel.

LATE HOWES

Turkey — Not Eatmor
Howes; at least 75% colored; count not over 120; fit for 20 days' travel.

Battleship — Not Eatmor
Howes; at least 85% colored; count from 120 to 140; fit for 20 days' travel.

Honker-Mistletoe — "Eatmor"
Howes; at least 85% colored; count not over 120; fit for 20 days' travel.

Holiday — "Eatmor"
Howes of uniform deep-red color; count not over 120; fit for 15 days' travel.

Pointer — "Eatmor"
Howes; at least 85% colored; graded over a ½-inch screen; count not over 90; fit for 20 days' travel.

Santa Claus — "Eatmor"
Howes of uniform deep-red color; graded over a ½-inch screen; count not over 90; fit for 15 days' travel.

OTHER VARIETIES

Paul Revere — Not Eatmor
Early Reds; at least 85% colored; count not over 110; fit for 10 days' travel

Puritan — "Eatmor"
Early Reds of uniform dark red color; count not over 90; fit for 7 days' travel.

Eagle — Not Eatmor
Mammoths (Batchelders, Hollistons); at least 85% colored; count not over 80; fit for 7 days' travel.

Iris — "Eatmor"
Mammoths of uniform dark color; count not over 75; fit for 7 days' travel.

Plymouth Rock — Not Eatmor
McFarlins; at least 75% colored; count not over 90; fit for 10 days' travel.

Pilgrim — "Eatmor"
McFarlins; at least 85% colored; count not over 75; fit for 10 days' travel.

Magnolia — "Eatmor"
McFarlins; uniform solid color; graded over a ½-inch screen; count not over 60; fit for 7 days' travel.

Faneuil Hall — Not Eatmor
Matthews; at least 85% colored; count not over 110; fit for 10 days' travel.

Bunker Hill — "Eatmor"
Matthews; at least 85% colored; count not over 95; fit for 10 days' travel.

Windmill — "Eatmor"
Matthews of uniform dark color; count not over 85; fit for 7 days' travel.

Blue Heron — "Eatmor"
Pointed Howes; uniform dark red color; count not over 100; fit for 10 days' travel.

Chipmunk — Not Eatmor
Smalleys; at least 75% colored; count not over 125; fit for 15 days' travel.

White House — "Eatmor"
Smalleys; at least 85% colored; count not over 120; fit for 15 days' travel.

Pocahontas — "Eatmor"
Smalleys; uniform deep red color; count not over 110; fit for 10 days' travel.

Dragon — Not Eatmor
Chipmans or Bugles; at least 85% colored; count not over 120; fit for 15 days' travel.

Samoset — "Eatmor"
Chipmans or Bugles; uniform dark color; count not over 120; fit for 15 days' travel.

OTHER VARIETIES

Myles Standish — "Eatmor"
Bugles or Chipmans; at least 85% colored; count not over 100; fit for 15 days' travel.

Beacon — "Eatmor"
Bugles or Chipmans; uniform dark color; count not over 100; fit for 10 days' travel.

John Alden — Not Eatmor
Centennials; at least 85% colored; count not over 90; fit for 7 days' travel.

Priscilla — "Eatmor"
Centennials of uniform dark color; graded over a ½-inch screen; count not over 75; fit for 7 days' travel.

Peacock — "Eatmor"
Centrevilles; at least 85% colored; count not over 90; fit for 10 days' travel.

Pheasant — "Eatmor"
Centrevilles of uniform dark color; count not over 75; fit for 7 days' travel.

Red Cedar — Not Eatmor
Black Veils; at least 85% colored; count not over 130; fit for 15 days' travel.

Lone Pine — "Eatmor"
Black Veils; well colored; count not over 110; fit for 10 days' travel.

Fisherman — "Eatmor"
Stanleys; at least 90% colored; count not over 110; fit for 7 days' travel.

Pioneer — Not Eatmor
Natives; at least 75% colored; count not over 120; fit for 15 days' travel.

***Inspected** — "Eatmor"
Any special variety having no established grade label; at least 85% colored; count not over 130; fit for 7 days' travel.

***Bluebird** — "Eatmor"
Any mixed variety averaging at least 85% colored; count not over 130; fit for 15 days' travel.

***"Variety"** — Not Eatmor
Any variety or mixed variety averaging at least two-thirds colored but having less than the established minimum grade color of any variety, containing not over 15% **all** white berries and 5% berries of a **green** color; count not over 175; containing not over 5% unsound berries. Name of variety must be plainly stamped or marked on the label.

Plainheads
Any berries which do not conform to any of the requirements of the preceding grades must be shipped in unbranded packages plainly marked as to size of package, variety, and packer number.

Pie Berries
Berries that pass through [illegible]-inch screen, regraded over a [illegible]-inch screen, and screened, having color of crop run of any variety, shall be marked —"No. 1 Pie Cranberries."
No berries will be accepted as Pie Berries which do not meet these requirements.

* These brands to be used only upon definite authority of an inspector.

New England Cranberry Sales Co. brand guide.
Middleborough Public Library

Enthused by the cooperative's success in Wisconsin, growers in New Jersey and in Massachusetts, including those belonging to the Cape Cod Cranberry Sales Company, formed their own cooperatives the next year, known as the New Jersey and New England Cranberry Sales Companies. Again, Arthur Chaney was their selling agent. Yet unlike Wisconsin, the lure of remaining independent in the marketplace died hard in Massachusetts and New Jersey, where only 35 and 30 percent, respectively, of all growers joined the sales companies at their inception. To this day, people knowledgeable about the cranberry industry find it hard to believe that a cooperative ever got off the ground in Massachusetts, considering the noted individuality of Cape and Plymouth County growers. That the cooperative idea overcame natural suspicion in the minds of the growers points up just how bad the market must have been in the first decade of the 1900s.

With state cooperatives in place in each of the country's major growing regions, growers took a further step toward cooperation and the delegation of responsibilities in 1907 when the three sales companies created the National Fruit Exchange, based in New York City, with Chaney as its general manager. Within four years, the Exchange went through a metamorphosis, merging with the old amalgam of producers known as the Growers' Cranberry Company and finally becoming the American Cranberry Exchange. It was the duty of the state sales companies to ensure that their members grew, screened, and packed the best cranberries possible. The responsibilities of the Exchange began after the berries were loaded in freight cars. It arranged for both transporting and warehousing the crop, but most importantly, for its advertising and sale. In placing its berries, the Exchange was able to bypass the commission merchants, who took a 7 percent commission on cranberries they sold to retailers. The co-op worked through jobbers who, for a small margin, brought cranberries and other fruit and produce directly to grocers for purchase. In the 1920s, when chain stores such as A&P were developing, the Exchange sold fruit directly to the store's own buyers. Both the American Cranberry Exchange and the state cooperatives were financed by a 7 percent assessment on the selling price of each barrel of berries. Five percent kept the operation of the Exchange afloat, while 2 percent was returned to each local co-op.

As a cooperative, the American Cranberry Exchange operated in an enlightened and democratic fashion, a fact not lost on members when a rival co-op with different principles evolved in 1930. Each state cooperative was represented on ACE's board of directors, according to the amount of acreage the co-op controlled. And most importantly, each co-op member, whether he grew four acres of cranberries or 400 acres, had a single vote in the affairs of both his co-op and the Exchange.

The energy that made the American Cranberry Exchange run successfully belonged to the capable Arthur Chaney. One of his first acts was to give the cranberry, previously sold ungraded and unsorted, an identity and a character. Between 1907 and 1910, Chaney worked out a detailed branding system for cranberries, which ensured that a wholesaler or retailer could order a brand of cranberries preferred by his customers and expect to receive those very berries, year in and year out. The brands defined the fruit by its variety, size, and keeping quality. The cranberry would no longer be a generic fruit. Perhaps more important, Chaney was committed to selling absolutely the best fruit possible. He reasoned that if people always found good cranberries on the market, they would buy them consistently; if not, they would simply buy another kind of fruit.

In what was perhaps an overzealous effort to target as many specific markets as possible and maintain the highest standards of quality, Arthur Chaney devised more than ninety brands of cranberries. His "Rules For Branding," as the key to cranberry brands was titled, was precise and particular and enforced to the letter by the state cooperatives. For a single cranberry variety, it defined a number of brands, based on the degree to which the berries were colored; their size, determined by how many would fit in a standardized cup measure; and how far they could travel and still remain fresh. Therefore, Peacock Brand cranberries consisted of the following: Centervilles (a variety of cranberries); at least 85 percent colored; count not over 90; fit for 10 days' travel. Pheasant Brand cranberries differed only slightly: Centervilles of uniform dark color; count not over 75; fit for 7 days' travel. Chipmunk Brand were Smalleys; at least 75 percent colored; count not over 125; fit for 15 days' travel, and White House Brand contained Smalleys only a bit darker and larger: at least 85 percent colored; count not over 120; fit for 15 days' travel. On and on went Chaney's litany of cranberry brands and their variations.

Packing the crop, Tremont screenhouse, Wareham, Massachusetts.
Middleborough Public Library

In those days, you could trace practically every shipment of berries right back to the bog it came from.

—Robert Hammond

Just before a Massachusetts grower shipped his crop, an inspector from the New England Cranberry Sales Company, usually a fellow grower, arrived at his screenhouse to determine what brands his berries would fall into. The co-op paid an incentive for the very best fruit, so each grower tried to get a considerable portion of his crop determined as the highest-quality brand. Growing the best berries meant more money—it was as simple as that. Beyond noting color and size, the inspector gathered data on the probable life expectancy of the fruit. He observed whether the crop was tender, spongy, damp, or sticky—all factors that would limit its lifespan. He asked if the cranberries had come from a bog fertilized or sanded in the past year, if they had been picked with scoops or snaps, and how long it had taken to screen them, questions calculated to learn whether the berries would hold up for a week or a month. In short, the inspector recorded the life history of these cranberries in excruciating detail. The quality of information gathered was so good, that as former cooperative member Robert Hammond remembered:

> In those days, you could trace practically every shipment of berries right back to the bog it came from. There wasn't any question about where it came from.... They knew where the better cranberries came from and the berries that they could ship to the West Coast and the stuff they'd better get into New York and get rid of quick and that sort of thing.

The variety of cranberry purchased mattered far more to the cooperatives than to the wholesalers, retailers, and customers, who could hardly tell the difference, but what did matter to them was the berry's color and how well it kept. After the system of brands had been operating a while, the Cranberry Exchange found that customers in different parts of the country liked cranberries of different shades. Westerners preferred pinkish fruit and considered deeply colored berries overripe, while customers in Baltimore and New York demanded the darkest cranberries possible. Both the sellers and consumers of cranberries were pleased to get fruit that did not turn to mush after two weeks on the shelf or in the icebox. Although somewhat arcane and overly detailed, Arthur Chaney's branding system brought a high degree of quality control to what had been a simple and old-fashioned method of selling the crop. The newly adopted brands gave the American Cranberry Exchange a peerless reputation as a cooperative committed to providing its customers with the finest fresh fruit.

The most beguiling thing about the brands of the New England Cranberry Sales Company were the brand names. Whether these were the products of Mr. Chaney's mind alone is unknown, but they played well with the cranberry's traditional association with historic New England, the Thanksgiving holiday, and by extension, the Pilgrims. When the Exchange ceased shipping berries in hundred-pound barrels and switched to quarter-barrel boxes, the names became more evocative, for the

American Cranberry Exchange box labels.
Middleborough Public Library

brands were illustrated on paper labels pasted to every box. There was Samoset Brand, and on the label, Samoset, the Maine Indian who first welcomed the Pilgrims to America, stood with arms outstretched. There was a Mayflower Brand, Myles Standish Brand, and John Alden and Priscilla Brands, John writing at a desk, Priscilla spinning at the wheel. Other brands honored New England institutions and landmarks, like Harvard and Yale Brands, and Minot's Light Brand. Another group evoked the region's flora and fauna with brands called Red Cedar, Lone Pine, Pheasant, and Honker; Canada geese migrate through the bogs in the spring and fall, and on the label, honkers beat their wings in an autumn sky.

The cooperatives in Wisconsin and New Jersey claimed their own brands and labels. In the upper Midwest, one could find Badger Brand cranberries and in the Pine Barrens, Tiger Brand or Quail Brand, the tiger being native to New Jersey only at Princeton, the bobwhite quail, more widely dispersed. Today, the brand labels of the American Cranberry Exchange are antiques and more likely to be found in a frame than on a box. They are admired for their illustrations and muted colors—dun, quiet browns, and reds.

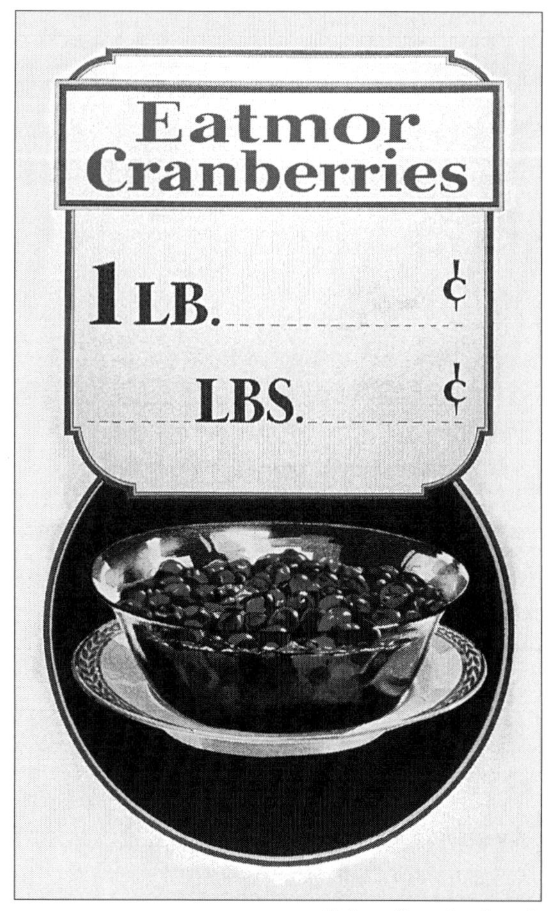

Eatmor Crannberries ad.
Private collection

Beyond his ubiquitous brands, Arthur Chaney could claim two more early accomplishments for the cooperative: success at setting an opening price for cranberries and in regulating their distribution. Where the commission merchant once decided what he would pay the grower for his berries, the co-op now determined the opening figure, a far better thing in the eyes of the members. Chaney's sources of pricing information were market surveys, which he ran frequently. While it is not easy to imagine all the factors that might affect the sale of the cranberry, Chaney thought he knew. Over the years, his list of the important economic indicators grew to include the following: the price of grain, which he perceived as being a general measure of prosperity; the Federal Reserve indices; the cost of sugar (since many women bought cranberries to can at home); the number of canning jars and canning jar rings purchased during the summer; and the prevailing ethnic mix of the population (since northern Europeans were familiar with cranberries and ate them, while southern Europeans were not and did not).

These diverse factors, combined with Chaney's feel for the market, plus the opinions of selected merchants, were the equation that equaled the opening price. In the end it was a tricky thing. If the price was set higher than wholesale and retail demand warranted, it dropped almost immediately, since the buyers determined the value of cranberries once the season began. The decline in price was often difficult to halt, and sometimes growers got no more for their fruit than they had spent in growing and harvesting it. But if Chaney had calculated correctly, and market demand justified the price, cranberrymen enjoyed a profitable year.

Orderly distribution of the crop also helped to stabilize the grower's income. Thanks to cooperative marketing, Philadelphia would no longer swim in cranberries at Thanksgiving time, while Detroit suffered a severe drought. Because the American Cranberry Exchange held the cranberries for many growers, it could ensure that merchants in every city received the fruit they needed, as they needed it. Theoretically, at least, this led to high demand for cranberries and good prices. Yet, since a large number of Massachusetts and New Jersey growers remained independent of the cooperative, if they assaulted a market at the same time as the Exchange, a cranberry glut resulted and prices plummeted. This very scenario was described by co-op members in attempts to persuade independents to join them. For the more cranberries the cooperative controlled, the more stable the market would become and profits to growers would rise.

Despite the innovations of Arthur Chaney for the new cranberry cooperative, things were rocky in its early years. Though it was possible to control some things, like the opening price and distribution, other events happened of their own accord. A string of good growing years from 1909 to 1914 produced larger and larger crops, while demand grew not at all, so prices remained the same as they had before the cooperative was established. Frost was still a major problem, often damaging the crop and forcing the co-op to put cranberries of lesser quality and appeal on the market.

While the weather was something that Arthur Chaney could do nothing about, he could influence another variable—customer demand. In 1918 the American Cranberry Exchange did something that had never been done before, it publicized the lowly cranberry nationally. When asked by the advertising firm just what his goal was, Chaney responded, "To get people to eat more cranberries." His dictum became the new brand name of the cooperative's fruit. Eatmor cranberries joined the dozens of other phonetical brands that sought the consumer's attention at the grocery store.

While the wholesaler or retailer continued to order Honker or Pheasant Brands specifically, the homemaker now knew these only as Eatmor cranberries. The name not only succinctly expressed Mr. Chaney's wish, but his marketing philosophy as well: he believed it was more likely that people who already ate cranberries could be persuaded to eat more, than that those unfamiliar with the fruit could be convinced to eat any. Accordingly, the ad campaign focused on locales where cranberry sales were already high. In November of that first year, advertisements in eighty newspapers across the country commanded citizens to eat more Eatmor cranberries. The message was also spread through the holiday issues of *Good Housekeeping* and *Ladies' Home Journal.* The advertising suggested new ways to serve the cranberry, and which sweeteners could be substituted for refined sugar, which was scarce during the First World War. To the great benefit of the growers, A. U. Chaney brought the cranberry, traditional fruit of American holiday meals, into the competitive food market of the twentieth century.

Of all the states to which the Exchange shipped cranberries, one proved most difficult to crack from the standpoint of marketing, and that was Massachusetts. This was so because of the "independents," those growers who resisted the cooperative impulse and went their own way. These men continued to sell cranberries, often under their own brands, in the time-honored fashion—through brokers, commission merchants, or handlers. While in most places Eatmor cranberries were the only cranberries in town, within Massachusetts, fruit of many brands and from many growers competed for attention.

In 1923 the American Cranberry Exchange, through the New England Cranberry Sales Company, controlled 65 percent of the Massachusetts crop and claimed the membership of nearly all the large growers. Those who remained independent tended to be the small growers, particularly those on Cape Cod, who feared anonymity in merging with an organization as large as the Exchange. Other reasons kept growers from joining, and one was expense. The 7 percent assessment that the Exchange charged its members to maintain operations was considered too high by some. The alternative for many growers was to sell through the Beaton Distributing Agency of Wareham, which started up in the 1920s and at one time, marketed as much as a quarter of the Massachusetts crop. Gibby Beaton, who worked for his uncle John in the family business, recalled why many chose the Beaton Agency over the Exchange:

> Of course, there was the New England Cranberry Sales Company, which was the big competitor. And their expenses were perhaps not exorbitant, but they were high. And he [John Beaton] was handling berries on the basis of twenty-five cents a barrel. He'd sell your berries for twenty-five cents a barrel. And his average return was most likely a dollar a barrel more than the American Cranberry Exchange was paying, which interested growers who don't like somebody who is big. Now there was nothing wrong with the American Cranberry Exchange, the Eatmor Company, they were a good, sound company, but they were big. And so there were some growers that, because they were big, said, "Well, we'd rather deal with a small person."

Not only how much money was earned, but how soon it was received, loomed large in the minds of some cranberry growers. The cooperative gave its members a partial payment on shipment, but growers waited until the season was over for the rest of their money. The most wary and conservative of the independent growers wanted cash on the cranberry barrelhead and sold their crop outright to a wholesaler. Others, in selling cranberries to a small commission merchant, assumed they would get their returns more quickly than cooperative members.

Perhaps the truest individuals among the independents were those who packed their cranberries under their own brand and label. These men wanted the best fruit they could possibly grow to have one identity: their own. Eino Harju of South Carver packed his own Pride of Carver Brand, and C. A. Ricker of Duxbury sent Land-of-Bays Brand cranberries to market. In Wareham, Handy and Hennessey stuck their Eagle Holt Brand label on the barrel or box, and on that label, an eagle fed its young cranberries, in a nest high above a bog. In Harwichport, William R. Wheeler fell prey to racism and branded his berries Piccaninnies, the label showing three tiny black girls dreaming of roast turkey and cranberry sauce.

Now there was nothing wrong with the American Cranberry Exchange, the Eatmor Company, they were a good, sound company, but they were big. And so there were some growers that, because they were big, said, "Well, we'd rather deal with a small person."

—Gibby Beaton

Wholesalers branded the cranberries they sold, too, and the charm of the brands lives on in the labels. The Colley Distributing Agency of Plymouth pictured two young Colleys in sailor suits on its label. Enjoying cranberry jam spread on bread, the youngsters were the best possible advertisement for Suitsus Brand. The venerable Boston commission merchants Hall & Cole sold the crop of many Cape growers and named their brand Kings Highway, after the old road along Cape Cod. Perhaps the labels that best expressed the importance felt by the growing region as the nation's largest supplier of cranberries were those of the Beaton Distributing Agency. The Long Distance Brand label pictured steam engines pouring forth from Wareham, hauling freight cars of cranberries to the far corners of the nation, San Francisco and Spokane. Beaton's Cape Cod Cranberries label showed the earth, and above it, berries pouring from a box. Massachusetts did literally cover the world with cranberries.

The image of the Massachusetts cranberry grower as independent, suspicious, casting a cold eye on groups and associations, is legendary. It once prompted Lewis Webster, an acting state commissioner of agriculture, to remark:

Independent brand box labels.
Middleborough Public Library & private collections

United Cape Cod Cranberry Company screenhouse, Hanson, Massachusetts.
Ocean Spray Cranberries, Inc.

I have always been very interested in the cranberry growers. They always seem to me to be the last outpost of rugged individualism. I remember meeting one of the growers to whom the government wanted to give a check for $350.00 for sanding his bog. He said that he didn't want to take money he hadn't earned, that he was going to put sand on the bog anyway, and he didn't want the government to pay him for doing it.

But coming along in years was an individual who would do all in his power to force the grower to shed long-held traditions. His headlong plunge into the business would enliven the cranberry industry for decades to come. On a day late in the 1800s, a boy from Sullivan, Maine, named Marcus Libby Urann visited a cranberry bog in neighboring Franklin with his mother. The Welch and Patton Bog was well kept, the accompanying farm looked prosperous, and then and there the boy told his mother that he would someday own a cranberry bog.

By 1910, Marcus Urann was both a young Boston lawyer and president of the United Cape Cod Cranberry Company, which soon controlled 600 acres of bog in Plymouth County. Urann had quickly proved himself an innovator and a force to be reckoned with in the cranberry industry. In 1907 he had helped to organize the American Cranberry

I have always been very interested in the cranberry growers. They always seem to me to be the last outpost of rugged individualism

—Lewis Webster, Acting Massachusetts Commissioner of Agriculture

Exchange, and during 1910 he planned and oversaw construction of a central screen and packing house, which was by far the most modern in the region and served all of his company's bogs. But in 1912 Marcus Urann did something particularly progressive: he began to cook cranberries in enormous steel kettles at his Hanson packing house and sell canned cranberries under the brand name Ocean Spray. Unlike the very direct message in the name Eatmor, the appeal of this brand was far more subtle. In the mind of the customer, it was meant to suggest an image of Cape Cod's lovely shore and of salt spray wafting over the nearby cranberry bogs.

Several companies toyed with dehydrating cranberries early in the 1900s, but canning was Urann's province alone. That Urann saw the logic in canning cranberries was not peculiar considering his Maine boyhood, for commercial canning began there in the mid-nineteenth century. Canned was the traditional way to sell Maine crops like tender wild blueberries, which otherwise perished before reaching market. Marcus Urann's new twist was to take a fruit that kept well and was traditionally sold fresh and put it into a can. He reasoned that canned fruit would extend the cranberry season to the entire year. Instead of buying fresh cranberries for sauce only at Thanksgiving and Christmas, consumers could now buy canned sauce throughout the year, increasing consumption and the market. Most cranberrymen thought Urann's idea was nonsense. Arthur Chaney, manager of the American Cranberry Exchange, tried to dissuade Urann from his plan; he saw no future in the canning business. The common belief of the day was that cranberries were a seasonal crop, mostly purchased for holiday meals. They had always been sold as fresh fruit and they would continue to be sold as fresh fruit. That was that.

From there, things went along much as they always had. Fresh cranberries continued to carry the market. Marcus Urann, while selling much of his crop through the cooperative as fresh fruit, persevered in canning some cranberries and in keeping the Ocean Spray Brand before the public. That Urann's idea was by no means outlandish became apparent when, in the 1920s, two more canners came on the scene. The A. D. Makepeace Company of Wareham, the nation's largest grower, now sold its own cranberry sauce through the Makepeace Preserving Company, and in New Egypt, New Jersey, the Cranberry Products Company canned berries from the Pine Barrens. Quickly, the competition became intense for what must have been a small market. To gain a larger share of the business, the price of sauce was continually cut by each company, until there was no longer any profit in it.

As the first cranberry canner, and the one that had established the best-known brand name through a vigorous advertising campaign, Marcus Urann's Ocean Spray Preserving Company could have outlasted the competition, but it would have taken time. Because Urann was not a patient man, he proposed something else: a merger of the three concerns. Although the idea was agreeable to John C. Makepeace of Makepeace Preserving Company and to Elizabeth Lee from Cranberry Products, Inc., of New Jersey, it would not have been agreeable to the federal government. Once combined, the canners would have held 95 percent of the cranberry sauce market, a clear and illegal monopoly. Thus thwarted, Marcus Urann commanded his legal counsel, "All right, then figure out what we can do that will get the same results and not be illegal." Rummaging around in a law library, the lawyer came upon a relatively new piece of legislation, the Capper-Volsted Act of 1922. The act allowed for the formation of agricultural cooperatives which included not only individuals as members, but companies as well. It was a law that seemed made to order for the three sauce makers, and in 1930, the cooperative Cranberry Canners, Inc., came into existence.

Once the cooperative was established, Marcus Urann took up its cause—grower control of the market, orderly distribution, and fair prices—as if they were the first words he had ever spoken. He called on every cranberry grower to join Cranberry Canners, and all he required was this: a pledge of 10 percent of the grower's crop. Urann reasoned that this percentage of every cranberry crop was unsuitable for sale as fresh fruit; there were always some berries that had been touched by frost or bruised by picking. But these cranberries still made fine cranberry sauce, and by sending them to Canners, the grower got both an outlet and money for fruit he could not sell fresh. The first members of Cranberry Canners, Inc., were also members of the American Cranberry Exchange: indeed, both Urann and Makepeace were on the Exchange's board of directors. The theory was that the cooperatives were not competitive, but complementary, since Canners only took fruit that the Exchange found unacceptable.

While many growers welcomed the new option, some remained wary of the proposition. They felt that canning was ignoble and defamed a fruit that for generations had been placed before the customer fresh from the bog and equal to every scrutiny. If a portion of their crop was not good enough for the fresh market, then it was not worth selling. These growers would have gladly died before any of their cranberries found their way into tin cans.

There was one aspect of the canning cooperative that many growers especially disliked, and that was the matter of voting power. Within the New England Cranberry Sales Company, the Massachusetts branch of the fresh fruit cooperative, the rule was "one man, one vote"—and the growers cherished it. But Cranberry Canners, Inc., was a stock cooperative. It needed more capital than its sister co-op because making cranberry sauce required cooking and canning plants and freezers. The capital came from its member-shareholders, and the amount they invested determined their voting power within the cooperative. Not unexpectedly, Marcus Urann's United Cape Cod Cranberry Company and the A. D. Makepeace Company controlled the majority of the stock and therefore, the cooperative. This arrangement was not what the growers were used to, and they didn't like it.

Like Arthur Chaney at the American Cranberry Exchange, Marcus Urann was the dynamo behind Cranberry Canners. In canning sauce, he looked to a day that most could not imagine, a day when women would leave the kitchen and do their work in offices. He saw an age approaching when convenient, prepared foods would be the norm, and he had a product waiting and his own slogan, "Ready to Serve," for when it did. "Change was a fact of life for Mr. Urann," said a grower who knew him. "In fact, he made change. He didn't just wait for it, he made it." Urann was also an inspired sales-

Ocean Spray Cranberry Sauce billboard and bog, Hanson, Massachusetts.
Ocean Spray Cranberries, Inc.

man, an old-fashioned huckster with boundless enthusiasm for his work. The birth of his new cooperative inspired him to write this patriotic address to the growers:

> Work as you must, worry as you will, kill bugs and flow for frost, still your profits depend on the supply and demand for cranberries.... Let us Cape Codders throw out our chest, take pride, and every day boost and blow for Ocean Spray Brand Cranberry Sauce. Ten million people will visit Cape Cod this year and they shall not pass without seeing, feeling, hearing, and tasting cranberries.

A cranberry grower's sister who worked in Marcus Urann's household said he was always playing around the stove and cooking up cranberries in one way or another, always thinking and experimenting. In the early 1930s, he expanded the product line of whole and jellied sauce to include

---

Let us Cape Codders throw out our chest, take pride, and every day boost and blow for Ocean Spray Brand Cranberry Sauce. Ten million people will visit Cape Cod this year and they shall not pass without seeing, feeling, hearing, and tasting cranberries.

—Marcus Urann

The Bottle, Onset, Massachusetts.
UMass Cranberry Station

> ***If he could get you to join Ocean Spray by just asking you to, he would prefer to do it in some other way, so he could say, "Well, I hooked him, got him in." And I liked Marcus Urann, he and I got along fine, but he was a promoter, a conniver.***
>
> —Gibby Beaton

cranberry juice cocktail. Sold in a tall and graceful bottle, it was marketed as a sort of medicinal drink that combated faintness and exhaustion as well as thirst. Most importantly, it sold cranberries. Some of Mr. Urann's store of energy was expended in the frequent use of axioms and proverbial sayings. "For every need there is a remedy," he said, an appropriate maxim for a marketing man, and "There are two classes of people, lifters and leaners." He was the former.

Marcus Urann's Achilles' heel was a tendency toward wiliness, a fact that did his business no harm but often made him suspect to Yankee cranberry growers. It is not hard to find stories that place Urann in a less-than-admirable light. Gibby Beaton of Wareham, who in his career worked for his Uncle John's Beaton Distributing Agency as well as for each of the cooperatives, offered several tales:

> They called the Makepeace, the Beaton, the Hammond [Companies] the Three Bigs who furnished berries to Marcus Urann, without any idea of getting anything back from them, to try to promote the process and get the business going. And John Beaton was one of the first directors of the canning company. That went along for quite a while, until Marcus Urann built the freezer in Onset, which was

Cranberry Canners' plant, Onset, Massachusetts.
Ocean Spray Cranberries, Inc.

at the time a cost of approximately two million dollars, and John Beaton called him up and says, "Well, Marcus, I don't remember our board of directors voting on this," and so forth. And Marcus says, "Well, John, you know you'd have voted for it if you'd been here—it didn't make any difference, just going ahead and building it." John resigned. He didn't think that was quite cricket. But we furnished cranberries, in fact, we gave them 10 percent of our crop and our growers' crop for processing, right through until Marcus Urann sent out an independent buyer to our growers and was buying berries behind our back to go to Ocean Spray.... Yes, I think there was a strong consensus of opinion that he was a promoter, he was an able promoter, and he did a tremendous job for the industry, but he liked to connive. If he could get you to join Ocean Spray by just asking you to, he would prefer to do it in some other way, so he could say, "Well, I hooked him, got him in." And I liked Marcus Urann, he and I got along fine, but he was a promoter, a conniver.

Ironically it was Urann's early training as a lawyer that seemed to allow him a certain freedom in his business practices. This was the feeling of the North Carver grower Larry Cole, who knew Marcus Urann and his cooperative very well.

Well, I got the sense that he had a great deal of confidence in himself and what he was doing, and that if he couldn't achieve something through the front door, he'd go around the back door. Again, he knew the law and he knew how far he could stray from the law and get by with it.

Within the cooperative, there was an ethical foil to Mr. Urann's business exuberances: John C. Makepeace. Mr. Makepeace was quiet, reserved, careful, scrupulous, and possessed an agile mind himself. As colleagues for twenty-five years, he and Marcus Urann checked and counter-checked each other, operating on principles that were often poles apart. Larry Cole remembered this, too.

Tosterettes and cranberry sauce ad. Private collection

I think that Mr. Makepeace's ability was as great as Mr. Urann's ability, one as a banker, the other as a promoter. And I'm sure that Mr. Urann kept out of a lot of trouble because Mr. Makepeace wouldn't let him do things that he probably wanted to do. Mr. Makepeace was strictly ethical in every sense of the word. He crossed his Ts and he dotted his Is and his corners were square, and it was just the reverse with Mr. Urann. Mr. Urann wasn't concerned about technicalities.

But even Urann's many detractors found it hard not to like him. He was something of a rogue—there was no doubt about it—but a very clever and charming rogue.

Hesitantly, the New England Cranberry Sales Company joined Urann's Cranberry Canners in 1934, pledging 10 percent of its members' berries, while still having little confidence in the business. It was not until 1937 that the Sales Company and the parent American Cranberry Exchange discovered that there was a certain value in canning. In that year the cranberry crop was very large and demand was low. The Exchange had a surplus of 200,000 barrels of fruit, which were fated to rot for lack of a market. Then, in rode Marcus Urann on a white horse. Borrowing one million dollars, Urann bought the surplus for the fair price of nine dollars a barrel, froze it, and sold it all as sauce over the course of the next two years. In particular situations, the fresh fruit partisans had to admit, canning cranberries had its advantages.

Mr. Urann did not expect his customers to use cranberry sauce only as a complement to the occasional turkey. His master plan was a year-round market for the fruit, and in advertising, he promoted a raft of new uses for the sauce. Spread it like peanut butter on Toasterettes crackers and you had a snack that relieved faintness, made the appetite keen, and even tasted good. It could serve as a filling for tarts and turnovers, jelly rolls, pies, and in cakes. Diced, it took the place of fresh cranberries in muffins and bread. Cut in a large cube, one could pretend it was a fruit salad or gelatin dessert, when topped with whipped cream. For those, and only those, who "persist in home cooking," Urann grudgingly suggested in one advertisement the purchase of Eatmor Brand fresh cranberries.

Compared to the creative flights of Marcus Urann, the American Cranberry Exchange took a much more traditional approach to its promotions. The cranberry, as everyone knew, was the fruit of Thanksgiving and Christmas. For housewives who canned fresh cranberries for later use, the Exchange offered no new recipes, only the information that cranberries were good for you. Touting a product's medicinal qualities was still a common advertising tactic in the 1930s. Eatmor Cranberries were styled as "the tonic fruit, rich in iron, lime, and carbohydrates—the vital elements that aid in restoring nerves and toning up the system." But it did not stop there. The Exchange also hailed the cranberry as a preventative of spring fever. This condition, an ad explained, occurred when the body's supply of calcium was depleted, and physical fatigue set in. Rich in calcium, the cranberry was just the ticket to keep spring fever at bay. Whether this approach to advertising was the ticket to increase sales was a different matter.

The decade 1940 to 1950 was a tumultuous one for America and for the cranberry cooperatives as well. As the country went to war early in the decade, so the cooperatives did in its later years, though the only comparison between the two events was the simple fact that, in each case, irrevocable conflicts existed. For the co-ops, those conflicts mounted steadily over the years. In 1938 the New Jersey branch of the fresh fruit cooperative, Growers' Cranberry Company, had joined Urann's Cranberry Canners, Inc., further increasing its pool of canning berries. In 1940 Marcus Urann and some fellow Massachusetts growers traveled to Wisconsin and won the agreement of the Wisconsin Cranberry Sales Company to pledge 10 percent of its berries to the canning co-op. A year later, Urann packed his bags for the West Coast and signed up the small growers of Washington and Oregon, cranberrymen whom the American Cranberry Exchange had never sought out for cooperative membership. In the space of eleven years, Cranberry Canners had garnered the participation of cranberry growers nationwide, many of whom it shared with the American Cranberry Exchange, and had built canneries in each of the growing regions. Its presence no longer seemed quite so complementary to the fresh fruit business.

In one other way, 1941 was a landmark for the cranberry industry. In that year, the Department of Justice brought an indictment against Cranberry Canners, Inc., the American Cranberry Exchange and its three state companies, Marcus Urann's United Cape Cod Cranberry Company, and the A. D. Makepeace Company—all the players in the cooperative game—for antitrust violations. The Justice Department accused the cooperatives of suppressing competition by independents in all manner of ways. The co-ops had, it was asserted, purchased cranberries from nonmembers to keep them from competitors, prevented the publication of advertising by the independents, sought to prevent the construction of canning factories by independent canners, and circulated false rumors about the credit of independent canners. It was an unhappy case in which the noble and shining five-syllable word "cooperative" had been reduced to the four syllables of a seamy and tarnished word: "monopoly."

A. D. Makepeace Company's evaporated cranberries. A. D. Makepeace Co.

There had long been monopolistic tendencies within the cranberry industry, in part because the industry was so small. In 1927 a consortium of New York financiers had sought to buy all of the most productive cranberry acreage in Massachusetts, New Jersey, and Wisconsin and thereby control the world supply of the crop. The plan did not transpire, but in few other agricultural enterprises was it even possible to conceive of one company owning almost all the means of production for a crop. Within the cooperatives themselves, the number of prominent growers was so small that many served on the board of directors of both businesses. This situation, if not smacking of monopoly, was at least highly incestuous.

From the standpoint of business, the war years were good years for cranberry growers. Despite the rationing of sugar and the shortage of tin cans, cranberrymen made money because they had a new customer: the military. From 1942 to 1945, one third to one half the cranberry crop was sold to the armed services in dehydrated form. The American Cranberry Exchange was selling vast quantities of fresh fruit to Cranberry Canners, who were removing the moisture and sending the government brick upon brick of pure, dried cranberries. During that time, the A. D. Makepeace Company stepped outside the cooperative arrangements and sold its own dehydrated fruit called "Crannies" directly to the military. Despite the call for America to pull together as a nation and overcome threats to freedom in the world, the branches of the service maintained their petty differences throughout the war. So remembers Ernie Howes, who worked in the Makepeace "evap" plant.

They evaporated the whole crop for the army and the navy. Now this is the interesting thing. The army had to have sliced berries, so we had to make a cutter to slice the berries. Then the navy wanted whole berries, so they had to make rolls with these gramophone needles and they punched the berries and dried them in that way. So you see? The difference between the army and the navy. They had to be different. So it made quite a problem. They overcome it.

Perhaps the most remarkable thing about the wartime sales was that they existed at all. Cranberries were not potatoes and could easily have been overlooked but for their important symbolic value. For this reason the army designated the cranberry an "essential crop," making it easier for cranberrymen to get fuel and supplies needed for growing fruit in that time of severe rationing. In 1944 the army alone bought thirty-five million pounds of turkey and a million pounds of cranberries for the Thanksgiving and Christmas dinners of servicemen. For the soldier lucky enough to get it, a traditional holiday meal of turkey and cranberry sauce must have lifted the spirits somewhat and served as a simple reminder of home.

During Word War II, no matter which way the grower sold his berries, he made money. The military, wanting sauce for the troops, paid good prices for the fruit. The housewife, in competition with army and navy procurement officers for cranberries, was willing to pay a premium as well. The average price per barrel of cranberries went up and up: $12.41 in 1941, $13.48 in 1942, $18.72 in 1943, and $26.08 in 1944. One other thing increased during the conflict, and that was the amount of fruit canned or dehydrated. The war had heavily favored cranberry processors and they took a larger percentage of the crop each season. A little more than one-fourth of the crop was processed in 1941, but by 1944 the amount was up to nearly half. The processors included not only Cranberry Canners, Inc., and the Makepeaces, but also the independent canners whom the cooperatives had tried to shut out of the business in the late 1930s.

World War II ended in 1945 just as cranberry growers were preparing to pick their crop. The removal of the military as a major purchaser of cranberries would bring an inevitable decline in growers' profits, but the effects weren't felt immediately. The public would again comprise the bulk of the cranberry market, but this time, half would be sold fresh and half would be sold in cans. Gone were the days when the cannery took only the small percentage of the fruit too poor to be sold fresh; Americans were becoming quite used to eating prepared foods that came out of cans, and that included cranberries. From now on, the fresh fruit marketers and the canners—the American Cranberry Exchange and Cranberry Canners especially—were in direct competition for each and every cranberry grown. And while the Exchange's membership was dropping, and business was still conducted as it had been in 1907, Cranberry Canners was expanding its business, product line, and membership. It now had 662 grower-members, while the Exchange could claim only 459.

Marcus Urann's perception of the die-hard fresh fruit grower and the American Cranberry Exchange as slaves to tradition—selling the same old product in the same old way—would eventually be depicted in a radio skit by the New England humorists Bob and Ray.

And this is radio's highly regarded Wally Ballou in another man-on-the-street broadcast, in our series of interviews with interesting people from all over this country of ours. I wonder if I could chat with you, sir, for just a minute.

I'd be more than delighted. It's a pleasure to be talking to a radio microphone.

Well, you don't get the chance very often, that's for sure. What's your name, please?

My name is Ward Smith. I'm from Ellisville, Massachusetts.

Isn't that down on Cape Cod somewhere, Ellisville?

Yes, just at the tip of…the start of Cape Cod.

That's beautiful country down that way. What do you do, sir?

Well, I thought you might guess. I'm a cranberry grower and I own cranberry bogs.

Is that so? I've always wanted to know a little bit about the raising of cranberries. They're such beautiful things when you see them growing.

Well, they grow on bogs, you know, Mr. Ballou. And, uh....

Have to be very careful of frosts don't you?

That's right, you have to flood the bogs if there's a danger of frost. Then you have to harvest them when they're big and red and ripe and juicy and bitter....

You know they're bitter?

Yes, of course. After you harvest them....

You harvest them in the fall, I guess, don't you?

Yes, late summer, early fall.

After you do that, you turn them over for processing. Do you process them yourself?

What do you mean by that, Mr. Ballou?

By that I mean, do you have your own factory for squeezing the juice out of them?

Squeezing the juice out of cranberries?

Yes, to make cranberry juice. Such as you find in cranberry juice cocktail, that sort of thing.

Well, you've triggered me a little. I've never thought of that.

Well, that would be a good thing to do with it.

Well, I've been selling them out of a basket, like strawberries, cranberry shortcake and that. They.... They really don't move that way.

Well, you could also sell them to people who wanted to make jelly out of them. Cranberry jelly is delicious.

Jelly out of cra.... Well, is there pectin in them or something?

Well, I suppose there is. And you can put certain additives in them. Cranberry sauce is very delicious, with whole cranberries in it.

Why, uh, wh.... Well, what would that be, a dessert?

Ye.... No. People serve it with things like turkey and certain kinds of meat.

Well, I've ah.... Well, ah....

There are a great many uses for cranberries, sir, that I think maybe you haven't stumbled upon as yet....

To Urann, selling only fresh cranberries in a grocery store was about the same as selling them out of a basket at the roadside.

There was an extremely large crop and extremely high returns to the grower in 1946; demand remained out ahead of supply. But processors handled more cranberries than did the Exchange, whose members feared that they were now losing control of the market they had successfully dominated for so long. Also in that year, the two cooperatives discussed a proposed merger, acting on the general belief among all growers that the co-ops should work in concert and that the alternative would spell disaster for everyone. While Cranberry Canners approved the merger, the Exchange turned it down. They may have believed that the co-op could withstand the incursions of Canners and return the fresh cranberry to prominence, or that the tyrannical Marcus Urann would never share power in a new cooperative, no matter what he agreed to. Rather than be pushed aside, they would take their chances in the marketplace.

With the merger effort a failure, Cranberry Canners, Inc., began to play rough and confirm fears that Urann was out to dominate the cranberry market. Canners broadened its name and scope in

The year I bought it, the first crop I picked,
I got twenty-four dollars [a barrel] for.
The second year I got fourteen dollars,
and the third year I got seven dollars,
and I think the fourth year or fifth year
I got three and a half dollars.

—Wilho Harju

Ocean Spray fresh fruit box label.
Middleborough Public Library

1946, becoming the National Cranberry Association and selling fresh as well as canned cranberries. Seeing the handwriting on the wall, the New England Cranberry Sales Company sold its stock in the National Cranberry Association in the following year and sent some cranberries to independent canners instead. Demand and prices remained high through 1947, and then the effects of severe competition began to show.

The scenario of decline was perhaps most clearly seen by young men returned from the service and just entering the cranberry business. One of these was Wilho Harju, who harvested his first crop on his first bog in 1947.

> The year I bought it, the first crop I picked, I got twenty-four dollars [a barrel] for. The second year I got fourteen dollars, and the third year I got seven dollars, and I think the fourth year or fifth year I got three and a half dollars. So you were almost cutting in half every year, and I figured, well, there wasn't much you could cut in half any more, when it got down to three, three and a half.

Not only was the fresh berry competing against the canned berry, but with the canner's fresh cranberries, too. A long and damaging price war that pitted the American Cranberry Exchange against the new National Cranberry Association ensued in every market in the country. Each co-op sought to undercut the other and so gain the lion's share of the market. But it was truly the cranberry growers who paid the price, for the low cost of cranberries in the stores meant small returns to growers. In some years, the proceeds did not cover the cost of raising the crop.

It was in the postwar years, too, that the specter of the surplus arose. In 1937, when Marcus Urann bought up surplus cranberries, froze them, and sold them over the course of two years, the plan seemed a panacea to the occasional problem of overproduction. But when advancing technology and cultivation techniques made overproduction the norm, the surplus cranberries became an albatross around the industry's neck. Despite the protests of growers, Marcus Urann persisted in holding over surplus berries from one year to another. He would not see a single cranberry rot if he thought he had even the remotest chance of selling it.

In 1948 the carry-over from 1947 was half a million barrels; in 1949 the 1948 carry-over was almost that much. In a market already plagued by low prices, these surpluses did nothing but produce a colossal glut.

To escape the ever-mounting supply of cranberries, for which there were few buyers, Marcus Urann and his advertising staff stepped up their promotional efforts in some subtle and not-so-subtle ways. Certain markets had always been good for the cranberry industry, specifically the cities of the upper Midwest. There, large numbers of Germans and Scandinavians bought and used cranberries year-round. In their native lands, the lingonberry—the wild European cranberry—was commonly eaten, so that in America, these immigrants were predisposed to buy cranberries and sauce. The fruit always sold well in places like Chicago, Minneapolis, Grand Rapids, and Milwaukee. But the cranberry industry wanted and needed strong markets nationwide.

Since the late 1930s, the canning co-op had operated the Cranberry Kitchen, where home economists experimented with every possible means of including cranberries in the American diet. They produced many dishes which were perfectly presentable, such as an array of fruit salads in which the cranberry played a role. Yet over the years, the culinary artisans in the Cranberry Kitchen also concocted some food combinations which were simply outlandish and suggested an almost desperate attempt to move cranberry sauce off the store shelves. The question remains: was the market increased or decreased by the invention of recipes such as tuna cranberry sandwiches, tuna cranberry omelets, cranberry butterscotch muffins, cranberry meatloaf, and liver loaf topped with cranberry sauce?

The origin of the deadly surplus, 1948, was the year in which cranberry sauce was promoted with absolute abandon. The National Cranberry Association came up with a new advertising tactic. They reasoned—somewhat belatedly—that if consumers were conditioned to eating turkey with cranberry sauce, they could be made to serve it with another kind of poultry which they ate far more often: chicken. The Chicken 'n Cranberry campaign was born, and joined a host of parallel promotions. A leading dress designer showed his wardrobe in the color cranberry red, and the cosmetic industry offered the same shade of lipstick for that season. The Arthur Murray Studios would soon introduce a new dance to their students: the Cranberry Bounce, consisting of a series of bouncing hops modeled after the journey of a cranberry through the separator.

The climax of these efforts came in late October when Governor Bradford declared a National Cranberry Week and the first Cranberry Festival was held in South Carver. A young Charlton Heston was flown up from New York and crowned the Cranberry King. Marcus Urann gave him a Chicken 'n Cranberry tie for his efforts. The Cranberry Queen, a local beauty, wore a crown with cranberries on the spike ends. Everyone sang the Cranberry Doxology. But the real event of the day was a wedding, the crowning achievement of the Chicken 'n Cranberry campaign. Before a clergyman, a larger-than-life white hen and a can of cranberry sauce were joined in matrimony and promised to appear together on dinner plates across the nation. A poultry farmer from Maine was the father of the bride, and John Quarles, the National Cranberry Association's legal counsel, stood up with the can of sauce. This wonderfully corny marriage of fowl and fruit was the sort of promotion the late 1940s and 1950s were famous for. But to beat the cranberry surplus, the industry would have to do better than this.

The Chicken 'n Cranberry wedding.
Ocean Spray Cranberries, Inc.

Although the early postwar years were an era of low profit or no profit for the cooperatives and their members, history was on the side of Urann's National Cranberry Association. More and more, growers like Larry Cole changed their cooperative allegiance to NCA, acknowledging the trend toward prepared foods.

Well, I switched over to the National Cranberry Association because I could see that there was going to be a change in the way of the consumption of cranberries. And again, I'm going to use a discussion I had with my father. My father used to remind me, "Larry, look how good fresh fruit has been to the industry." I said, "Pa, when you and I go to the store, we buy what we want in the form we want it, if it's available. And who am I to say how our customers

---

"Pa, when you and I go to the store, we buy what we want in the form we want it, if it's available. And who am I to say how our customers are going to buy cranberries? If they want it out of a can, that's their judgment. If they want it out of a glass, that's their judgment. I don't care. They're going to buy it as they want it."

—Larry Cole

National Cranberry Week.
UMass Cranberry Station

are going to buy cranberries? If they want it out of a can, that's their judgment. If they want it out of a glass, that's their judgment. I don't care. They're going to buy it as they want it."

For other cranberrymen the advantages that the canning co-op now seemed to possess brought about an awful realization, one best expressed by Wareham grower Robert Hammond:

When you could take a poorer quality product—not poorer quality in the sense of what it was going to be used for, if it was going to be used for processing, because it would make just as good a sauce, but poorer quality in terms of fresh fruit, which had been the predominant marketing method for years. When you could take the poorer quality berry and put it into a tin can and get more money for it, it didn't seem right. But what do you do when it's a matter of economics, and they could do it? It was a fact of life.

All of a sudden the American Cranberry Exchange's unwavering dedication to selling the finest fresh fruit meant little. The dozens of brands, thousands of visits by cranberry inspectors, the intricate system of quality checks on which the co-op's reputation was based had been eclipsed by simply putting the cranberry in a can. And with the decline of absolute quality as the industry standard came another loss: the incentive for growers to raise the best cranberries possible. Simply put, quality wasn't so important anymore. The withering of the American Cranberry Exchange in the face of competition by the National Cranberry Association was a hard fact for many to accept. Of the two cooperatives, ACE was the sentimental favorite. It was the first cooperative (many of its members were the sons of members), it had served the industry well for a long time, and it was the most democratic, giving equal weight to the interests of large and small growers.

Back on the cooperative battlefield, both sides sat down once again to settle their differences. In 1949 the New England Cranberry Sales Company proposed a marketing program that would serve the interests of both co-ops and end the cranberry cold war. The plan called for the distribution of all fresh fruit by the American Cranberry Exchange and processed fruit by the National Cranberry Association. All co-op members would hold membership in each cooperative, and the entire program would be overseen by a governing board known as the Cranberry Growers' Council. The proposal was the obvious and sensible solution, and both the directors of ACE and NCA approved it unanimously. But the plan was thwarted, according to Larry Cole, who worked diligently in its behalf, by one substantial stumbling block: Marcus Urann.

When you could take the poorer quality berry and put it into a tin can and get more money for it, it didn't seem right. But what do you do when it's a matter of economics, and they could do it? It was a fact of life.

—Robert Hammond

Even though openly Mr. Urann indicated that he wanted the Cranberry Growers' Council, behind the scenes he did all that he could to frustrate it so it wouldn't succeed, because he knew eventually that the Sales Company would collapse, and when it did, he—Ocean Spray—would be the one organization.

In 1954 Marcus Urann's wish came true. The New England Cranberry Sales Company, in substantial debt and beset by creditors, sold its assets to the National Cranberry Association. It was a victim in the expensive competition with NCA for growers and cranberries. With the largest of the sales companies eliminated, it would not be long before the American Cranberry Exchange (which had recently become Eatmor Cranberries, Inc., to trade on its brand name) would fall, too.

Marcus Urann retired as the president of the National Cranberry Association in 1955 at the age of eighty: "Rest is rust," he always said. The agenda that many felt he had had for so long—to control the entire cranberry industry—was largely completed. In 1957 Eatmor Cranberries, Inc., the national fresh fruit cooperative, went the way of its Massachusetts branch and dissolved. And then there was one.

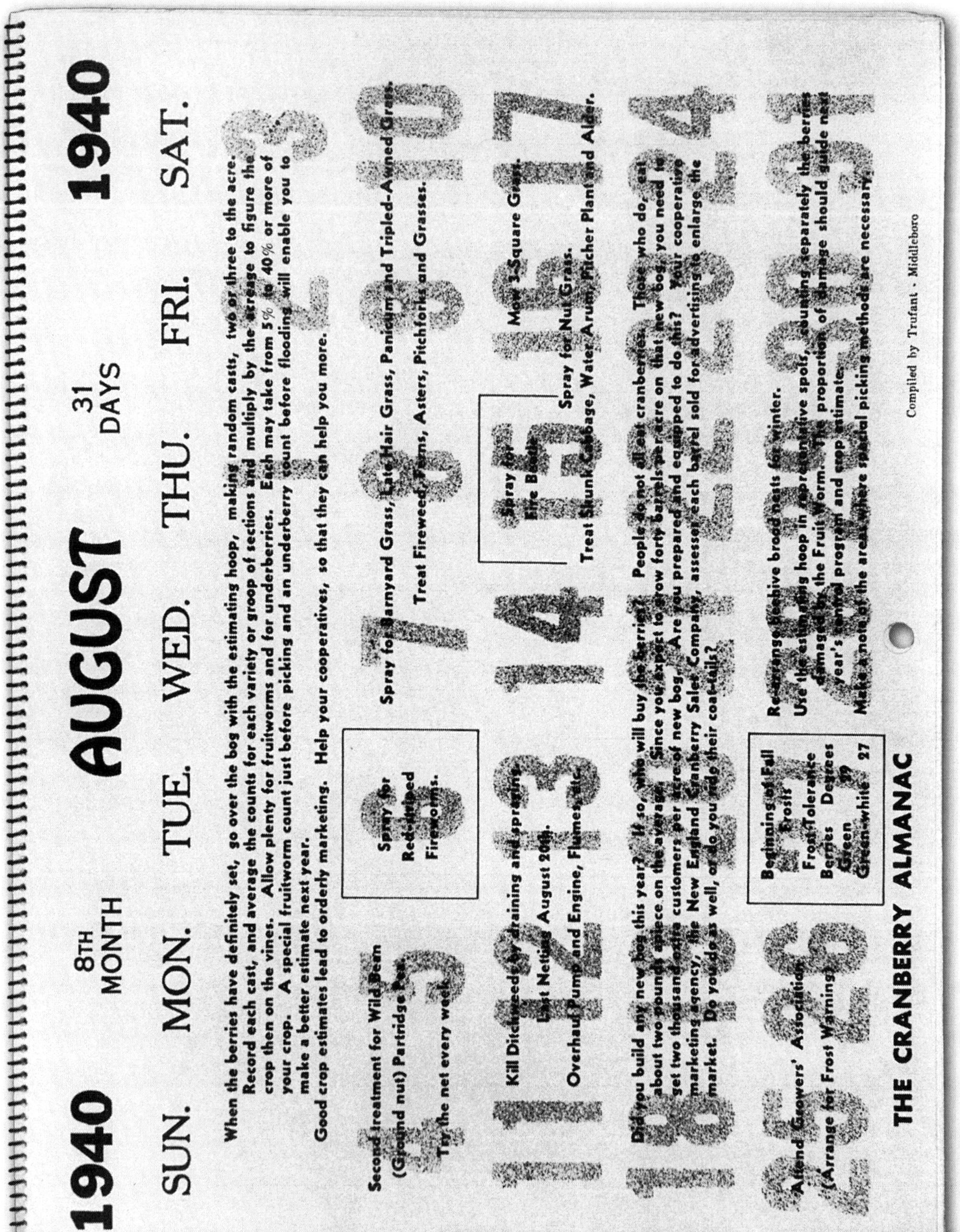

1940 8TH MONTH **AUGUST** 31 DAYS **1940**

SUN. MON. TUE. WED. THU. FRI. SAT.

When the berries have definitely set, go over the bog with the estimating hoop, making random casts, two or three to the acre. Record each cast, and average the counts for each variety or groop of sections and multiply by the acreage to figure the crop then on the vines. Allow plenty for fruitworms and for underberries. Each may take from 5% to 40% or more of your crop. A special fruitworm count just before picking and an underberry count before flooding will enable you to make a better estimate next year.

Good crop estimates lead to orderly marketing. Help your cooperatives, so that they can help you more.

1 2 3

4 5 6 7 8 9 10

Second treatment for Wild Bean (Ground nut) Partridge Pea.

Try the net every week.

Spray for Red-striped Fireworms.

Spray for Barnyard Grass, Late Hair Grass, Panicum and Tripled-Awned Grass.

Treat Fireweed, Ferns, Asters, Pitchforks and Grasses.

11 12 13 14 15 16 17

Kill Ditchweeds by draining and spraying.

Last Netting August 20th.

Overhaul Pump and Engine, Flumes, etc.

Spray for Fire Beetle

Mow 3-Square Grass.

Spray for Nut Grass.

Treat Skunk Cabbage, Water Arum, Pitcher Plant and Alder.

18 19 20 21 22 23 24

Did you build any new bog this year? If so, who will buy the berries? People do not all eat cranberries. Those who do, eat about two pounds apiece on the average. Since you expect to grow forty barrels per acre on that new bog, you need to get two thousand extra customers per acre of new bog. Are you prepared and equipped to do this? Your cooperative marketing agency, the New England Cranberry Sales Company, assesses each barrel sold for advertising to enlarge the market. Do you do as well, or do you ride their coat-tails?

25 26 27 28 29 30 31

Attend Growers' Association.

(Arrange for Frost Warning)

Beginning of Fall Frosts

Frost Tolerance

Berries	Degrees
Green	[illegible]
Green-white	27

Re-arrange Beehive brood nests for winter.

Use the estimating hoop in representative spots, counting separately the berries damaged by the Fruit Worm. The proportion of damage should guide next year's control program and crop estimate.

Make a note of the areas where special picking methods are necessary.

THE CRANBERRY ALMANAC

Compiled by Trufant - Middleboro

The Cranberry Almanac.
Middleborough Public Library

Chapter 10

The Grower's Almanac

In the year 1940 a grower from Middleboro named Russell Trufant compiled for fellow cranberrymen a calendar and almanac of the year's activities. It chronicled for every month and season the necessary work to be done on and around the bog—work that had changed little in method and execution since the start of the century. Trufant prepared the almanac with the small grower in mind, the fellow who had six to ten acres of bog and did much of the work himself without benefit of hired labor. In style, the almanac emulated the column called "Farmer's Calendar" from that primer on New England agrarian life, the *Old Farmer's Almanac.* It was a work as clever as its maker, Mr. Trufant, and here and there, gave evidence of the understated humor for which he was known.

JANUARY–FEBRUARY

Cutting logs for shipping boxes for future crops. Nailing up next year's shipping boxes.

Until the 1920s, the Massachusetts cranberry growers sent their crops to market in hundred-pound barrels—the common containers for so many commodities of the day. But because cranberries are perishable, the barrel was not the ideal receptacle from which to sell them. Unless a grocer sold a lot of cranberries quickly, the last pound of fruit wasn't nearly so good as the first had been. To alleviate the problem of rotten berries at the bottom of the barrel, some growers hit on the idea of shipping in a smaller container. First came the fifty-pound half-barrel box and soon after, the quarter-barrel box, holding twenty-five pounds of cranberries. It would remain the standard container for shipping and selling fresh cranberries until the advent of cardboard and cellophane in the early 1950s.

Few growers handled the full range of box-making operations themselves, but one that did was A. D. Makepeace Company. In 1925 it put up its own box mill at Tihonet Village, Wareham, and soon thereafter, a sawmill to cut the boards from which the boxes were made. In winter, the Makepeace crews worked in the woods felling white pine and cutting the logs into 64-, 58-, and 54-inch box lengths suitable for handling in the mill. These logs were trucked or carted to the sawmill where the sawyer and his help worked them up into box boards. Outside the mill, in an area known as the board field, box boards by the thousands were stacked to dry in the air until the following winter.

The board field at A. D. Makepeace Company.
A. D. Makepeace Co.

The A. D. Makepeace Company box mill, Tihonet Village, Wareham, Massachusetts.
Lindy Gifford photograph

In the box mill, the native pine boards were reduced to box parts collectively known as "shook" and then made up into boxes in a process with nine individual steps. Through a building filled with complex, dangerous, and screeching machinery, the boards ran an industrial gauntlet. First it was to the planer, which planed them to a thickness of one inch. Next to the gale matcher where the bark was removed, the board sawn, a tongue and groove made, and glue laid on the groove. The squeezer squeezed the tongue-and-groove boards together, making a double shook. The gang-saw divided the double shook down to two single shooks and the re-saw cut the shook lengthwise. At the double-saw slotter, ventilation slots were cut in the shook for the box sides and bottom. The cleat ripper produced four cleats at a time for the four corners of the box, and the printing press stamped "Makepeace" and the current year in black ink on the box ends. Finally, at the rear of the mill, the nailing machines nailed the bottom and sides together, producing the cranberry box. The box mill turned out not only quarter-barrel shipping boxes, but also the larger harvest boxes, which pickers filled with cranberries on the bogs.

Most cranberry growers—far smaller than the Makepeace Company—left all or nearly all the box-making labor to the region's mills, which thrived by supplying the cranberry trade. One of the best-known box makers was Frank Cole of North Carver, but he was far from alone, as his son Larry made clear.

Well, there were two in Carver, Jesse Holmes and my father. There was one in Rock Village, Middleboro, L. O. Atwood. And there was Gilbert West in Pembroke—there's bogs over there—and Lot Phillips is a little more removed. And there was Acushnet Sawmills in New Bedford. So there was competition, but my father was in a good place and the competition didn't bother my father one bit.

Sign and machinery, A. D. Makepeace Company box mill.
Lindy Gifford photograph

My father would probably employ thirty-five people year in and year out. Now, in the fall of the year, we'd employ double that number and when there was a good crop the mill would start the first of September with a night shift, and that night shift would be working until the Christmas market.

—Larry Cole

Bog in winter, Carver, Massachusetts.
Lindy Gifford photograph

Work in the Cole box mill never ceased, but did vary in intensity from season to season.

My father would probably employ thirty-five people year in and year out. Now, in the fall of the year, we'd employ double that number and when there was a good crop the mill would start the first of September with a night shift, and that night shift would be working until the Christmas market. And before then, in the wintertime, he'd make shooks. Well now, you take the cranberry grower, he'd have his steady workers and with a lot of snow on the ground before the days of mechanical ice sanding, what was he to do with them? So my father'd make the shooks and he would truck the shooks to that screenhouse and they would take their steady year-round help and when there was too much snow on the ground and they couldn't do anything else, they were there nailing boxes.

Sanding on the ice.
Ocean Spray Cranberries, Inc.

Clearing around the bog.

Winter months were also a good time for clearing trees and brush back from the bog. Given the choice, most growers preferred to work outside no matter the season, and clearing allowed for this. The slash produced from cutting made a good fire, which kept the grower warm. There were sound reasons for fighting the forest's encroachment. A cranberry bog closely ringed by woods is a cold pocket; opening the land up increases the air temperature by a few degrees and reduces the threat of frost. Clearing around the bog also shortens the period of morning shade, allowing picking to begin earlier on autumn days.

Sanding the bog on the ice.
Reading Florida postcards from all your neighbors.

With the right conditions, January and February found a thick layer of ice on bogs flooded to protect the vines from the rigors of winter. Ice gave the grower an opportunity to sand his bog, a task the conscientious man carried out every three or four years. For decades, cranberrymen had known that the relationship between sand and the cranberry vine was an important one, and that occasionally applying sand on a bog increased its yield, but no one knew why. By 1940 it was generally acknowledged that a thin layer of sand provided a fresh growing medium for vines, allowing runners to put down roots and revitalizing root-bound vines. Sanding on the ice-covered bogs meant that with a Model A truck, the grower could carry great loads of sand over the bog doing no damage to the vines. And it also meant that he avoided the hundreds of wheelbarrow trips from sandpile to bog by which sanding was accomplished in the spring or fall.

With the exception of this year's stock market, I can think of no way to drop fifty thousand dollars more quickly than to put it into a cranberry bog if you haven't the knack for growing cranberries.

—Dr. Neil Stevens

Swallow and birdhouse.
Lindy Gifford photograph

MARCH–APRIL

Your civic duties. The assessed value of cranberry bogs and their consequent burden of taxation gives every cranberry grower a lively interest in his government.

Late winter/early spring wasn't the busiest time of the year for the cranberry grower; he still had some weeks to plan for the growing season ahead and time to devote to other duties, such as town meeting. Within his community, the grower was often regarded as intelligent and capable, since he was engaged in a particularly complex branch of agriculture. Dr. Neil Stevens, a scientist who worked with both Massachusetts and Wisconsin growers, felt that "as a group cranberry growers seem to stand well above the average in intellectual ability and individuality. Much of this I believe to be due to the intrinsic difficulty of the cranberry business." In 1929 Stevens had written, "With the exception of this year's stock market, I can think of no way to drop fifty thousand dollars more quickly than to put it into a cranberry bog if you haven't the knack for growing cranberries."

In 1940 the grower was, more likely than not, a Yankee, perhaps the tenth or twelfth generation of his family to live in town. Sometimes he carried the mark of old roots in his first name: a name such as Elnathan or Seabury, Nahum or Ruel. He was the son and grandson of cranberry growers and it was from them that he learned how to tell if frost was likely and when to start the winter flood—the important details of his work that no agricultural school taught. It was on bogs that his family had built and worked where he now made a living. He loved the work. He was his own boss, and he labored outdoors with no one to bother him. Often it was not the only work he did. For years it was thought that a man needed ten acres of bog to support a family, and many small growers had less. So the cranberry grower would often be a carpenter or mechanic, as well. Perhaps he worked winters in a sawmill and managed a bog for a wealthy family, as well as cared for his own property.

The grower rarely ran the bog alone; it was usually a family enterprise. His wife stayed awake with him on frost nights, and tallied during picking season. Sons and daughters helped with weeding and the dozens of tasks that the harvest demanded. Everyone felt a responsibility to the bogs, the family's primary source of income. Beyond work, the cranberry grower took part in small-town life. He might be a deacon of the Congregational Church or a member of the local Republican Committee. Perhaps he served as town tree warden or as a selectman.

Prepare and place your birdhouses.

Around the bogs, there were some chores to do now that the winter flood had been drawn off and the dull red vines began to green up. Each year, the cranberry growers practiced what was known as economic ornithology, pitting the insects who sought to destroy the vines and fruit against their natural enemies—birds. By the time swallows returned in early May to nest, the grower had ready birdhouses which were attached to long poles sunk into the main ditch or along the shore. The swallows were a tradition on the bogs and brought him pleasure in several ways. They wheeled, twittered, and snatched bugs from the air above the bog, a wonderful diversion from such tedious work as early weeding or cleaning ditches. And the sight of the birds filled the cranberryman with satisfaction, for there was no thriftier way to rid the bog of pests.

Some chores were so automatic, so ingrained in the cranberry grower's consciousness, that formal prompting in an almanac was not required. The loathsome task called ditching was one of these. While weeds began their ascent up through the vines on the bog, aquatic plants prepared to choke the ditches with growth. Clogged ditches meant that water for frost protection and irrigation traveled less efficiently around the bog. Left unchecked, plant life would simply fill in the ditches. So the cranberryman donned waders, picked up a long-handled potato digger, and went to work. Standing in the narrow ditch, he dredged out algae and pickerel weed, dropping it in piles on the bog. He followed the highway of ditches through his bogs until these channels were clean, doing some one day, some the next. If you stretched the work out a little, it wasn't so hard on the arms and back. The job was only complete when many mounds of organic debris had been wheeled off the bog. Like death and taxes, the yearly round of ditching was all but inevitable.

Philip Brackett, on his endless rounds of ditching, Chopchaque Bog, Mashpee, Massachusetts.
Lindy Gifford photograph

Dr. Henry Franklin delivers the frost warning report.
UMass Cranberry Station

MAY—JUNE

Active frost season begins.

The real nemesis of every cranberryman was neither insects nor weeds, but frost. Living enemies of cultivation reduced the harvest by some degree each year, but only a killing frost could take the whole crop overnight. The danger of frost was imminent in two seasons: late spring and early fall. In spring the terminal bud was at risk, the nub at the end of the upright branch swelling into a flower from which the berry would come. The bud would withstand temperatures as low as 20 degrees until the end of April, but as it grew it became more tender, more susceptible to the freezing air that came suddenly as frost.

The grower's only warning of a coming frost had always been his own weather sense. He knew that a cold, dry wind blowing throughout the day boded ill for the bog at night. The wind would die down around dusk and cold air from a high pressure system centered over the cranberry district would flow into the sheltered lowlands which held the bogs. Conversely, a day of low, heavy clouds, if they persisted after dark, usually assured the grower a good night's sleep. Freezing air would not penetrate such a defense. Yet conditions could change quickly, and predicting the weather was a tricky business. Sometimes the grower awoke to find his bog just touched by the cold, or far, worse, entirely frosted and every berry taken. In those years he sought other work to make up for income lost in one cold night.

Efforts to see dangerous weather coming received a considerable boost in 1920 when Dr. Henry Franklin of the Cranberry Experiment Station introduced a frost warning service. Based on weather data he had been collecting in Massachusetts since 1912, Franklin devised a formula that determined the minimum nightly temperature on the bogs hours in advance of its occurrence. The formula was based on temperatures and dewpoints reported to Franklin from volunteer weather observers in central Massachusetts at Boylston and at Rockport on the North Shore. He added in the same information for Wareham, made an allowance for wind speed, and had the minimum bog temperature. It was an amazing achievement.

To get the warning out, Franklin established a telephone relay system and enlisted the aid of growers' wives and other friends of the industry. These women became frost warning distributors, responsible for contacting growers in their area who subscribed to the service and reporting the impending frost. In the afternoon and evening during frost season, Dr. Franklin took calls from his weather observers, checked the Wareham readings, and then made the calculations, usually with a few anxious growers looking over his shoulder. If a frosting for the cranberry bogs looked likely, the ladies of the frost warning service were alerted. In minutes, a housewife on the lower Cape was dialing Osterville 4644 to tell grower Malcolm Ryder at Wakeby Lake; while off-Cape, the phone rang at Middleboro 168-M-2 and bog owner Minnie Dunham was apprised of the coming weather. Many of the women remained remarkably loyal to the frost warning service, some calling in spring and fall for more than a quarter century, others passing the job on to daughters only after many years. The service did create one problem, termed "The promiscuous calling of the distributors by the growers" in an annual report of the growers' association. A distributor wrote, "Mr. ——— has found out my telephone number and he really does bother me terribly. There have been days when he has called me eight and ten times to see if there is going to be a report. I always tell him I will call him when there is a report but he still calls."

A vintage Bailey pump. Lindy Gifford photograph

On the bogs, preparations for frost followed a routine. In the afternoon, after the first frost report was received, growers placed planks in the outlet flumes, sealing the bogs. With a pump or by gravity flow they then filled the ditches with water from their reservoirs—and waited. Franklin's formula could not project the hour when frost would descend, so it was important to ready the bog early. For those growers who owned the many acres of dry bog still under cultivation in Massachusetts, the only hope was a prayer for deliverance. The final frost warning was issued at 8:00 P.M. It reported either that the threat had diminished, in which case the grower went to bed, or that Jack Frost was indeed planning a visit. With that news, men returned to the bogs and continued the flood. When the water level rose within several inches of the vine tops the flood was halted, for heat from the water passed into the air and kept the remaining exposed vine from freezing. The long hours flooding the bogs were lonely and very cold.

Philip Brackett sweeping for insects, Chopchaque Bog, Mashpee, Massachusetts. Lindy Gifford photograph

JULY–AUGUST

Try the net every week.

"You don't go dancing on frost nights or vacationing in the worm season." That's what caring for bogs meant to a grower from Cataumet named Robert Handy, and it was never truer than in the heart of summer. The region swelled with visitors in those months; ladies from Little Rock peered down at Plymouth Rock and sunbathers lay the entire length of Cape Cod's beaches. Everyone seemed to have time for some relaxation except the cranberry grower. In early July the bogs bloomed with thousands of minute pink flowers that gave the landscape the coloring of salmon. The blossoms were so thick that honeybees brought to the bogs by beekeepers visited forty million flowers in pollinating a single acre of cranberries. Later in the summer, the delicate cranberry honey the bees produced was sold along with the other regional specialty, beach plum jelly, at roadside stands.

In these months, the grower was part entomologist, part groundskeeper. He spent a lot of time—as he had throughout the spring—slaying insects and keeping the weeds at bay. To learn which insects were dining on the bog, he used an insect net that at a distance made him look like a lepidopterist in search of swallowtails. Once a week the conscientious cranberryman swept his bog. Walking a zig-zag pattern, he ran the net back and forth through the vines like a sailor swabbing the deck.

You don't go dancing on frost nights or vacationing in the worm season.

—Robert Handy

Dr. Franklin inspecting for insects.
Middleborough Public Library

He carefully examined and identified the worms and bugs trapped in the net, sometimes with the help of a magnifying glass. If the number of a particular insect was high, the grower took action.

The chief pests on the bog were these six: the root grub, the fruit worm, the blunt-nosed leafhopper, the black-headed fireworm, the gypsy moth, and the girdler. Growers had strategies for dealing with all of them. The most clever and sustainable was still another use for the water that protected cranberry plants from winterkill and frost. If a grower held the winter flood on the bog into late spring, he could often check infestations of fruit worm, which ate their way into cranberries, and gypsy moth caterpillars, which defoliated the vines. Some insects, like the root grub, were done in by reflowing the bog in late spring and keeping the water on into early summer. The damaging side effect of reflowing was that it sometimes injured or killed the blossoms, reducing the crop as much as the grubs. A Cape grower tells the story of how his father's bog was badly infested with root grubs one season during World War II when cranberry prices were spectacularly high. The Cranberry Experiment Station suggested the bog remain under water all summer and that year's crop be forgotten. The man did as he was told, incurring the wrath of his wife, who was furious at the loss of so much money. He almost didn't survive that season, but his bog did.

Second dust or spray for fruit worm.

The alternative to drowning insects or letting the swallows do the work was to kill them with sprays and dusts. The organic pesticides pyrethrum and rotenone were popular in those days, as were the virulent and effective poisons sodium cyanide and lead arsenate. Spraying a bog was unpleasant in the extreme. On the shore, the poison and water were mixed in a tank and then pumped by hand or power through hosing. On the bog, one or more men tramped around pulling the heavy hosing and spraying the vines with a hand-held rig. At day's end, their trousers might be soaked with pesticides, and, perhaps, their lungs. Many claim spraying had no ill effects, some think otherwise. Dusting was just beginning to be done with machines that resembled motorized bicycles. These contraptions were run back and forth across the bog, spewing

Spraying pesticides.
UMass Cranberry Station

poison from spreaders at the rear. Wind could be a great problem for these dusters, particularly on Nantucket, with its constant breeze. Marland Rounsville, who managed Milestone Bog on that island, devised a way to beat the wind. He mounted lights on the dusters and ran them at night when the air was still. But it was a victory achieved at a cost. Working at night, he and his men were almost eaten alive by mosquitoes.

Treat fireweed, ferns, asters, pitchforks, and grasses.

Weeds on the bog—competing with the vines for nourishment and space—brought the grower as much pain as insects, but most of it was back pain. The means for dealing with weeds were more direct than those for insects; generally, they were either cut down or pulled out. The less fussy man usually waded onto the bog and swung a scythe just above the vines. To him, that was weeding. The careful grower went to the heart of the matter. He pored over the bog on hands and knees in the summer sun with a weed hook, pulling out every living thing that didn't answer to the name *Vaccinium macrocarpon*. On the biggest bogs, weeding crews of as many as thirty men and women crawled along, pulling baskets behind them. Every year it seemed to be something new. One summer a rash of maple seedlings would be the problem; the next, the whole bog was abloom with asters. Some growers were remarkably fastidious about their bogs. To them, owning a weedy bog was akin to living in an unkempt house. On the Cape, one grower was said to carry a salt shaker on the bog. Whenever he came upon a weed, he bent down and seasoned it liberally.

Russell Trufant tosses the estimating hoop.
Ocean Spray Cranberries, Inc.

When the berries have definitely set, go over the bog with the estimating hoop.

By midsummer, the berries had set. Blossoms became little green ovals that looked surprisingly like grapes. Water was kept high in the ditches now, so if rain was scanty the growing fruit would not lack for moisture. From now on, it paid the grower to stay off the bog with spray rigs, dusters, and his feet. But there was one more task the *Almanac* suggested, and that was forecasting the crop with an estimating hoop. It's no surprise that this duty was included, since the estimating hoop's maker was the *Almanac*'s author, Russell Trufant.

The process was not magic, but a matter of arithmetic. It happens that there are as many cranberries in a hundred-pound barrel—the industry's standard measure—as there are square feet in an acre, roughly 43,500. The wooden hoop was one square foot in volume and to estimate his crop, the grower ranged over the bog tossing the hoop into the vines and counting the berries within its circumference. By averaging the count from several tosses, the farmer learned what his yield per acre would be. Many growers felt that the set of berries on a bog was too uneven for the hoop to give a reasonable estimate; they preferred to eyeball the crop as had always been done. Their reservations didn't stop Mr. Trufant from believing in the accuracy of his hoop or from posing for an instructive photo. With an air of confident nonchalance, he demonstrated the proper technique when tossing the estimating hoop.

Bog, ditch, and shore.
Lindy Gifford photograph

Green frog in the ditch.
Lindy Gifford photograph

Perhaps now, in high summer when the world was most alive, the grower paused to admire what he had. It was much more than an island of plants set in sand and resting on peat, the waterlogged, compressed vegetation of the eons. It was a place of several distinct habitats which he shared with others. The bog itself was home to more insects than he cared to imagine and the ditches supported a complete food chain. There were diving beetles and water striders. Green and pickerel frogs dwelled both in the ditches and on their banks. Finning in the water were long and narrow pickerel, seemingly shaped by the confines of the space. The largest resident was also the grower's least favorite. Muskrats made their dens in dikes and undermined the bog edge in search of nutgrass and roots. Swimming in the ponds was the rarest and most wonderful creature of all, the Plymouth red-bellied turtle, who fed on vegetation and lived nowhere else in the world but Plymouth County, Massachusetts. In the air above, predatory dragonflies patrolled, including the rare banded bog skimmer. And higher, the diminutive wood warbler known as the northern parula shared the sky with swallows.

The land surrounding the bog was another entire ecosystem. It started with a grassy strip closest to the ditches, known as the shore. The term was old and recalled the mid-nineteenth century, when cranberry growing was starting up on Cape Cod. Many of its first investors and experimenters made their living on the sea and saw bogs in that light, though the coast of a cranberry bog cannot compare to the Cape's.

Out of the woods came the shore's animal life, most often at night. Raccoons and skunks searched the ditches for unwary frogs, their tracks visible the next morning. White-tailed deer browsed where the woods and shore met, then bedded down on the bog for the evening, leaving great, flat patches

Bogs, Plymouth, Massachusetts.
Lindy Gifford photograph

where they had lain. Some growers maintained that sharp deer hooves were nature's best seed planters, helping weeds get a good start on the bog. From the air, sparrow hawks scoured the land for voles and field mice, always in competition with red foxes who hunted below. Least shy of all were the Canada geese. They liked the shore and bog—whether flooded or dry—equally, sources of tasty vegetation. Cranberrymen complained that geese in numbers did a good job of fouling the bog, no pun intended. Some, particular about their bogs, threatened to bag more than the season limit of geese.

The bog was not only a shared place, but a thing as immutable as a house or cemetery. When a bog was built and planted with perennial vines, it became a permanent fixture of the landscape. There were Massachusetts bogs on which a crop had been taken for more than a hundred years and half-century-old bogs were common. It was not only from their longevity that cranberry bogs got an identity; it came also from their individual shapes, their peculiarities, and their names.

When nineteenth-century and early twentieth-century growers built a bog, they followed the contours of the land to the inch. A swamp's random shape became the bog's blueprint. And although the theory of the day said that small bogs were more profitable and easily worked than large bogs, they were built in all shapes and sizes. There were

Milestone Bog, Nantucket, Massachusetts.
Lindy Gifford photograph

Frog Foot Bog, Wareham, Massachusetts.
Lindy Gifford photograph

round bogs, oval bogs, some vaguely diamond shaped, others oblong, and bogs serpentine and wasp-waisted. For each variation in nature's design of low and wet land, there was a cranberry bog to match. The range in size was just as great, from pieces less than a quarter of an acre to a bog once 235 acres in size—Nantucket's Milestone Bog, touted as the world's largest.

Growers knew bogs in other ways, too. Each one had its own nature, they said, prone to wetness or frost, high at the edges so the winter flood did not cover it well, or a good producer year in and year out. Most symbolic of identity was a name. A bog was almost always named and took its name in one of many ways. The name of a bog builder or early owner often stayed with the property, becoming a sort of living genealogy. Thus, bogs like the Lawyer Lovell Bog, Hannah Bog, Smalley Bog, and Sally Jones's existed long after their namesakes departed.

Description, both poetic and prosaic, was also the origin of names, such as the Egg Bog, shaped in an oval. High Piece, Shoe Piece, and Holly Tree Bog. Place names accounted for some: Shaving Hill Bog, Wankinko, Swan Holt—and acreage for others: Fifteen Acre, Two Acre. Whimsy was not without its place in the cranberry business, at least on the bogs known as Duck's Dinner and Old Dandy. Nor was commemoration. One grower named a property the Johnny Bog to honor a grandson born the same year the bog was built.

Wheeling berries ashore.
University of Massachusetts Cranberry Experiment Station

SEPTEMBER–OCTOBER

PICKING TIME

The tail end of summer and the first weeks of fall were the climax of the grower's year—picking time. Once the fruit was well colored, the cranberryman was assured of working straight-out, nearly around the clock, until the cranberries were off the bog. If he was clever, he had long ago developed a good relationship with Cape Verdean men and women whom he counted on to return and scoop the crop each harvest season. When a grower's holdings were large, he managed the work of others, but if his acreage was small he was both overseer and worker. He might wheel berries off the bog himself while his wife tallied the pickings.

The longer the harvest continued, the harder the work became. Indian Summer days—hazy and overly warm—were hard on men and women who spent seven hours on their knees with backs bent. Squabbles erupted on the bog out of exhaustion. The owner was jittery, too, because picking time was also fall frost season. To come this far and lose all the year's income overnight was too horrible to contemplate. So he sat up on cool evenings, battling sleep and checking the thermometer. If the bogs had to be flooded, it put off picking for a full day while the vines dried out. The more days that berries stayed on the vines, the greater number of frosts they had to endure. Growers who raised both early and late varieties worked a longer harvest. After the Early Blacks were cleaned up, there would be a few days' rest while the Howes berries ripened. When these were well colored, the pickers returned to the bogs. They might not leave again until mid-October.

For all its rigors, the great human effort that was picking time is often fondly remembered by growers and pickers alike. Memories of sleepless nights, of aching muscles and blistered hands are replaced by those of long friendships and celebration at the harvest's end. Witness William E. Crowell's recollections:

> The proposition of hand-picking was very cheerful. You'd get maybe thirty or more people up there, they had a wonderful time, you know. They did a nice job, there was clean picking—they didn't waste much.

Picking time. Ocean Spray Crannberries Inc.

And they were very cheerful, very nice about it. When they got all through they'd have a big party in the screenhouse over at Locke Bog. And they'd cook up some chickens on the stove that was there and have a chicken barbecue and a couple of drinks and so forth, do a little dancing. That was part of the tradition—they enjoyed it. I did, too, I thought it was very nice.

It is sometimes permissible to take two days off during this period.

With the crop in, all breathed a sigh of relief. The berries sat in screenhouses—either the grower's or the state cooperative's—waiting to be shipped to market or the cannery. Some were screened out by skilled women and packed immediately, but most would not move until Thanksgiving or Christmas were nearer at hand. Now might be the right time for the cranberry grower to take those few days off and put his legs up. Read the newspaper once again and maybe sleep until seven in the morning. Take in a Grange supper, and to thank the wife for her help and forbearance, attend the Harvest Moon Ball at the Redmen's Hall.

This is also the season for raking the vines.

In tune with the rest of the landscape, the cranberry bog was transformed in autumn. Like the maples in the woods, its leaves turned a ruddy, rusty red with the approach of colder weather. It was in these newly colored environs that the grower did a little touch-up work on the bog as October ran out. The harvest always damaged the vines, scoops broke them off, wheelbarrows and feet twisted them this way and that. With a vine rake and a knife rake, the cranberryman coiffured the vines a little and put things right. Old, long vines produced few berries and were hard to push a scoop through. The knife rake took care of that. A rake with blades for teeth, it cut away ancient, tangled vines and left short, healthy ones. Newly shorn vines were then combed with the vine rake, which trained them all in the same direction as a further aid to scooping. With that, the year's outdoor work was pretty much complete.

Screening at Tremont screenhouse, Wareham, Massachusetts.
Middleborough Public Library

NOVEMBER–DECEMBER

November is another active month. Indoors, screening, packing, and shipping are in full swing.

As the holidays loomed at the end of the year, the rail yards in Wareham and Harwich and Falmouth and Bourne—in all the small towns of the growing region—grew busier, and truck traffic on the roads became heavier. Berries had been departing for distant points since the harvest's end, but with the two biggest turkey-and-cranberry days of the year just ahead, the pace became frenzied. At the screenhouse, women from town arrived early to screen as they had for many years. Talking of this and that, they kept their hands and eyes busy searching for the bad berry as thousands surged past on the belt. In the background was the low and constant "thunk" of cranberries ricocheting off the bouncing boards and pouring forth from the wooden separators—still more fruit to be inspected. The day was long and the building cold in late fall, not to speak of the facilities, which were rustic.

Facilities, Federal Furnace screenhouse, Carver, Massachusetts.
Lindy Gifford photograph

We had a load of quarter-barrel or half-barrel boxes ready for the freight. We'd meet Mr. Makepeace on the road and he'd be barrelin' down. He'd get off that road as fast as he could and let us go by.... 'Cause we had to meet the freight. The freight left Wareham at a certain time, destination New York or California, wherever...

—Ernie Howes

The grower carefully packed his cranberries, heaping them up in the box so they wouldn't slosh from side to side and bruise in transit. With the truck loaded, he headed for the station and the waiting freight cars of the New York, New Haven, and Hartford Railroad. These trips to meet the freight had priority over everything else within the A. D. Makepeace Company as Ernie Howes remembered it.

And you know, many times Mr. Makepeace would meet us on the road, we'd be coming down from Wankinko—that's out toward Plymouth. We had a load of quarter-barrel or half-barrel boxes ready for the freight. We'd meet Mr. Makepeace on the road and he'd be barrelin' down. He'd get off that road as fast as he could and let us go by. He would never hold us up. And many times he'd gone in a bad place where he was stuck and of course—Bill Ross

Shipping berries by rail.
Middleborough Public Library

at that time was the mechanic—he'd go out and pull him out. But he'd never hold us up, never. 'Cause we had to meet the freight. The freight left Wareham at a certain time, destination New York or California, wherever, and it had to leave at that time 'cause the engine was hooked up to that freight car and it had to move. So Mr. Makepeace says, "I'm not holding anybody up."

In 1937 alone, 190 freight cars left Wareham Center for the cranberry markets of the world.

Outdoors: sanding is the major activity, together with repairs to the dikes and flumes.

Having shipped most of his berries, the cranberry grower began looking around the bog for a final few chores that needed finishing up. If a man had not sanded in winter or early spring, fall was a good time, since there was no new growth to damage with the work. For the small grower, this meant wheeling more sand than he cared to think of over planking and spreading it by shovel on the vines. Many Yankee farmers hired Cape Verdean laborers to do this tiring work.

A large grower would use his own bog railroad to accomplish the task. At the start of the century, Russell Trufant and his father had pioneered the use of narrow-gauge railroads on Massachusetts

A narrow-gauge bog railroad.
Courtesy Eunice Bailey

cranberry bogs. They were a luxury only the wealthiest could afford. The narrow-gauge cars and rails were a railroad in miniature: the open cars were only as high as a man's waist, and the track—which came in detachable sections—was but two feet wide. On vast expanses of acreage, the railroad was the perfect means to move sand from the shore to the bog. At first the cars were pushed by hand, then pulled by horses, but by the 1930s an iron horse was available. Sitting on a miniature locomotive, a cranberry grower looked like he'd stumbled into a carnival. Indeed, the origin of the railroad theme park Edaville Railroad amid the bogs of South Carver was the narrow-gauge railroad used on Mr. Ellis D. Atwood's cranberry bogs.

The Winter Flood: applied generally in December, but sometimes unnecessary until January.

A long year of work at an end, all that remained was to flood the bog when the ground froze and the wind began to blow. The vines could stand the cold, but when the wind dried out the leaves, the frozen roots could not replace lost moisture and the plants died of dehydration. This was winterkill. With the bog under water for the next few months, the grower began to think about a Florida vacation for himself.

The scoop and the Darlington picking machine.
Ocean Spray Cranberries, Inc.

Chapter 11

Slow to the New

With the close of World War II, thousands of Americans returned home from every point on the globe to begin life again. The political landscape of the hemisphere would never be the same, nor would the lives of its citizens.

Making the trip with men and women to America were machines. A great deal of technology, first used in Europe, the Pacific, and the Mediterranean, was also demilitarized and came home to serve civilians. Like millions of others, the cranberry grower was suddenly in the midst of more change in his industry than he'd known in a lifetime. First of all, new equipment was available: bulldozers for building new bog and rebuilding old bog, power shovels for digging canals and reservoirs. A grower could hire an operator and his machine by the day or week, or buy his own 'dozer if his holdings were large enough. Airplanes and helicopters had a peacetime use for farmers, as well. Air Force aces and 'copter pilots now flew missions over cranberry bogs, applying chemicals far more efficiently than the grower with his pump and hose. Cutting a thirty-foot swath across the bog, a helicopter could dust twenty-five acres in an hour. The same work done by hand took days. And aerial application meant that buds or berries were not injured by men and machines on the vines. For Larry Cole, this new means of spraying and dusting was a decided improvement over the old method, but it still had drawbacks.

In 1940—and I guess I mentioned to you about the gypsy moth destroying that bog, that was before there were helicopters and aeroplanes—my uncle had bought an Arlington sprayer. It was a wooden tank that could hold a couple hundred gallons and you'd have a lot of hose and you go out and spray the bog. Well, that was better than what had been before but it still left a lot to be desired. Took a lot of help—you dragged the hose over the vines, and when the aeroplanes and helicopters come along, that left a lot to be desired, too.... You're on a list in the first place, so I call the man up and say I want my bog treated. Now, before I had sprinklers I was glad to get it done that way. You had to wait four or five days, but the insects, they didn't wait. Along come a rainy spell, you know. So around eleven o'clock one morning I was down the bog and the plane was spraying my bog and the wind was blowing it right off in the woods. And I called the fella up and I give him the devil. I said, "What you do it for?" "Well," he said, "you're on the list and you better take it today 'cause it's gonna rain tomorrow." I said, "The bog didn't get it, all blew up in the woods. It didn't do me any good." Obviously, I paid the bill, but it wasn't his fault either, he had such a long list.

Ford truck ad.
A. D. Makepeace Co.

The chemicals that planes and helicopters now spread were of a different order than the old organic pesticides pyrethrum and rotenone. These new compounds were synthesized in laboratories, the result of war-related research. At the head of the list was dichlorodiphenyltrichloroethane—DDT—a potent pesticide hailed as a boon to mankind which, while toxic to bees, was thought to harm wildlife not at all and affect aquatic life only minimally. There were many others. Chemicals called Aldrin, Chlordane, Dieldrin, Malathion, and Parathion all came on the market in the postwar years and were used by cranberry growers and every other agriculturist.

I was down the bog and the plane was spraying my bog and the wind was blowing it right off in the woods. And I called the fella up and I give him the devil. I said, "What you do it for?"

—Larry Cole

In these years, too, modern marketing dawned in America, and where the cranberry crop had once been merely packed, now it was pre-packaged, pre-weighed, and pre-priced. A wooden container of some sort had been the industry standard for shipping fruit since the beginning of commercial cranberry growing, but now the quarter-barrel box of native pine gave way to one-pound cellophane bags and cardboard boxes with plastic windows. In small groceries, cranberries had been sold right from the box—the clerk scooping out and weighing each purchase. But inside the new phenomenon known as the supermarket, consumers waited on themselves, choosing the berries that looked best through the cellophane. In the growing region, the move to cellophane and cardboard hurt the box mills badly. While continuing to make picking boxes, they lost the other half of their business in the decline of the shipping box. Larry Cole's father switched to building tonic cases in North Carver to recoup his losses, and Jesse Holmes & Son of Center Carver placed a sad and plaintive ad in *Cranberries* magazine. It said: "Remember Us?? We Are Still Manufacturing Cranberry Boxes."

In updated packaging, cranberries even left the district by a new means: over the road. Rail yards no longer hummed when fall came around, but the highways were busy. The evolving network of state roads was taking its toll on the trains. More cheaply than by freight train, trucks took cranberries to Boston on coastal Route 3, or traveled inland through Plympton, Whitman, and Abington on Route 58. With New York City as the destination, a trucker first drove across southeastern Massachusetts to Providence, Rhode Island, and headed south and west from there. In the few years between the beginning and end of the 1940s, the changes brought about in American life were extreme.

Fitted out with an agricultural arsenal that now included bulldozers, helicopters, power shovels, airplanes, DDT, and other postwar spinoffs, the cranberry grower enjoyed what one of his number called "the advantages of laziness"—the benefits that labor-saving equipment brought to cranberry growing. Efficiency on the bogs increased, except in several key areas. One was harvesting. A crop that was protected from insects by helicopters and packaged in cellophane was still picked with the tools used at the very start of the century: hand-made wooden scoops. While eager to take advantage of new technologies available to all, cranberry growers were slow to do something they had once done well: invent new tools to serve themselves when no one else would.

There are several reasons that help explain why cranberrymen entered the age of mechanized harvesting at such a leisurely pace. In the 1920s, 1930s, and through the end of World War II, profits from the bogs had been high and growers could afford the expense of picking the crop by hand. Nearby was an enormous pool of immigrant laborers whose lack of language and job skills made them almost a captive workforce for the cranberry industry. Cape Verdeans returned to the bogs every fall, like clockwork, to scoop cranberries. Hence, the growers' complacency and inability to realize that the future might be different from the past.

A limited effort to design and manufacture a power picker had been ongoing within the cranberry industry since the early 1920s, although it received little support among growers. The project had come about because of a slight labor shortage the industry felt during World War I, a shortage of the young white workers who supplemented the Cape Verdean help at harvest time. The problem made some men wonder whether such heavy reliance on hand labor was a healthy thing. In 1921 the New England Cranberry Sales Company, the

Massachusetts cooperative, scrutinized plans for pickers and some actual machines submitted by inventors—a few cranberry growers among them. They chose to underwrite the work of a young Finnish machinist from Quincy named Oscar Tervo.

Tervo's plan was to perfect a power picker that could be ridden, one that was literally the size and weight of a small car. By 1922 Tervo had produced a machine which the cooperative's picking machine committee found satisfactory, but not quite rugged enough. Back to the shop he went. When he emerged in 1923, the committee discovered that Tervo's machine performed far worse than it had before the improvements were made. At this point, Oscar Tervo, who had tinkered with his machine for months without success, walked away from it all and took an extended vacation. When he returned to Quincy, it was to work in the machine shop of a man named W. M. Mathewson. There Tervo decided to take up the recalcitrant machine once again, this time with help and financing from his new boss.

Through 1924, work on the mechanical picker went ahead. It was field tested, then adjusted, tested again, and adjusted again. Finally, in 1925, the power picking machine was pronounced a success and limited manufacture began. Despite the long years Oscar Tervo spent in a workshop ministering to his machine, it was not to bear his name. It was called the Mathewson picker.

What the Tervo-Mathewson combination had produced was a motorized behemoth. On a heavy steel tractor chassis sat an automobile engine, driver's seat, and steering wheel. The picker consisted of a revolving cylinder with fourteen rows of steel teeth attached. The teeth gleaned berries along a path two and a half feet wide and carried them to a movable apron. The apron shuttled cranberries to the other side of the machine, where they dropped into a box. Besides a driver, the Mathewson required another attendant who walked alongside and removed full boxes from the machine, replacing them with empty ones. Barring too many serious mechanical failures, it was estimated that one machine could pick two acres in a day.

Opinion today shades from those who would not let one on the job, through those who keep one "for insurance," and those who accept it as an additional tool, to those who use it for everything except brushing their teeth.

—Russell Trufant

Five of the larger and wealthier cranberry growers in the region took immediate delivery on the Mathewson power picker, and so began a long period of debate about its value that never really ceased. An early user, Middleboro grower Paul Thompson, declared that the Mathewson picked "better than a crowd of careless scoopers and I think better than the average scooper at piecework." This was somewhat more than faint praise. The machine also handled the berries carefully, leaving them in better condition, many thought, than those picked with scoops. Its major drawbacks were: reliability—it would break down as soon as not; its weight was nearly two tons; and it had a tendency to dig into the turf and make huge divots if the bogs were not absolutely 1evel.

Throughout the 1940s, the Mathewson picker controversy continued, and Russell Trufant reported that "Opinion today shades from those who would not let one on the job, through those who keep one 'for insurance,' and those who

The Mathewson picker.
Middleborough Public Library

accept it as an additional tool, to those who use it for everything except brushing their teeth." Trufant had himself purchased a Mathewson in the 1930s. "Ever since then," he revealed, "that machine has been ruling the bog with an iron hand, figuratively and literally."

Perhaps the fairest assessment of the often praised, often maligned Mathewson power picker came from William E. Crowell of South Dennis, who did not get his Mathewsons until the early 1950s, when the price of cranberries was so low, he could no longer afford to hand pick.

> The Mathewson picker was a mechanical nightmare. You went to work in the morning, spent about an hour and a half oiling it to start with, all these oil holes, you know. There was an awful lot of machinery to it…unless you had a mechanical mind, it was pretty hard to do much with it, frankly. Once you got it along—it was a big machine and it had a scoop that wide, you know—it would go right along and pick to beat the band. I guess you could pick eighty, ninety barrels a day with it without much difficulty. It had a four-cylinder Chevrolet engine on it—some of them had four-cylinder Fords, I think—but mine were Chevrolets, both of them. Of course the motor was running—you had it in second gear, it had a reg-

ular automobile transmission and engine in it. It weighed about 1,600 pounds, I think. It had a pair of tractor wheels in the front and a pair of wheels on the back, oh, about that wide. It was heavy and it was bulky and it was awkward to get up and down. We had oak planks we used to get it up and down the sections, and so forth.... On Early Black vines and some of the other vines that were crisscrossing like them, you'd lose a lot of berries. But on upright Howes vines and similar that stood up, it was almost perfect. It would do an almost perfect job, better than the modern pickers would. That's my observation. With the berries at the price they were, we could afford to lose a few and pick with a machine. Had to have an operator, and you had to have someone going alongside of it to take off the berries and so forth. But two men could do pretty well with it.

In the end, the Mathewson power picker was used only sporadically over the course of about twenty-five years. A good idea that had gone in the wrong direction, it was too big and lumbering, too complex, too temperamental, and too expensive. In an industry of many part-time growers with small bogs, the Mathewson was a luxury that only the more prosperous could afford. To some, the Mathewson machine was a hopeful sign that a suitable mechanized picker would eventually be developed; to many it was proof that the cranberry would always be harvested with a wooden scoop.

The next development in the search for an adequate picker followed a different path than that taken by Tervo and Mathewson. Taking a cue from the appliance revolution going on in households across America, an inventor in the state of Washington and one in Carver, Massachusetts, demonstrated picking machines inspired by the vacuum cleaner during the late 1940s. Transported from the living room carpet to a carpet of vines, the machines were simply outsized vacuums—large canisters on wheels with hoses attached—powered by gasoline engines. The Carver machine incorporated a scoop at the end of the hose; the berries were actually scooped by the operator and then sucked into the canister and emptied into boxes. The Washington machine employed three hoses accommodating three men, but used no attachments. Its aim was to vacuum the cranberries right from the vines. In reality, the machine's suction simply wasn't strong enough to accomplish this and the picker was forced to rub the berries off the vines in an exercise that one observer called "the hardest work in God's creation." Neither invention got far beyond the demonstration stage, suggesting that vacuum machines were better adapted to collecting dust in the home than cranberries on a bog.

While the vacuum pickers were under trial, a few men were putting the finishing touches on the machines that would harvest the crop for some years to come. The Western Picker invented by the Stankavich brothers of Coos Bay, Oregon, appeared on some Massachusetts bogs in 1948. Six years later, a picker designed by Thomas Darlington of the Whitesbog Cranberry Plantation in New Jersey, was harvesting Cape Cod cranberries. These machines had been prefigured nearly a hundred years earlier in two cranberry harvesters, one patented by Richard DeGray from the unlikely location of New Orleans, Louisiana, and the other patented by Joseph Jenney of Mattapoisett, Massachusetts. The old and new machines shared a basic wedge-shaped design, and were essentially two-wheeled harvesters guided through the vines by an operator walking behind. In each case, teeth captured the berries and a continuous apron or conveyor belt delivered them to the rear of the machine and a receptacle. Only one component kept a motorized cranberry harvester from operating in 1874 instead of 1948 and that was the small gasoline engine—a refinement not available until the 1900s.

Western Pickers.
Ocean Spray Cranberries, Inc.

Some growers have said flat-footedly that there could never be a picking machine which would be satisfactory. However, many are finding picking by the Western machine to have very definite advantages and not only in cost.

—Clarence Hall, *Cranberries* Magazine

Despite the belated presence of the mechanical cranberry picker in Massachusetts, a certain psychology made growers greet the machine coolly and with skepticism—the psychology of conservatism. In 1951 Clarence Hall, publisher and editor of the industry magazine *Cranberries,* titled an editorial "Are We Slow to the New?" In it he wrote:

Concerning cranberries we have been doing a little thinking as to the question, "Do cranberry growers like change as well as others? Have we made changes in cultural practices as rapidly as some other forms of agriculture have, or even as fast as may have been desirable?"

Hall went on to say,

We have our cranberry queens and our airplanes, our modern insecticides and our machine pickers. Yet the thought haunts us as to whether we are sufficiently alert to explore every new avenue which may lessen cost of production. Do we dismiss the possible advantages of the "new" too easily? Some growers have said flat-footedly that there could never be a picking machine which would be satisfactory. However, many are finding picking by the Western machine to have very definite advantages and not only in cost.

Maintaining a Darlington picker.
Lindy Gifford photograph

Growers' suspicion of the picking machine was the same expressed by men a half century earlier when wooden scoops came on the scene. They feared the invention would damage their bogs and bruise the cranberries. In fact, the mechanical harvester was less destructive to the bogs and fruit than the scoop. It was also faster and cheaper. A grower who first used the Western picker in 1949 estimated that he cut harvest costs by fifty cents per barrel of cranberries, using only six workers to pick thirty acres of bog. Scooping would have more than doubled the crew size.

> ***I remember the old lady down the street here, when they started with machines, she used to go watch them. And you'd see tears in her eyes.***
>
> —Doris Gomes

Picking machines were not without their problems, however. Like scoops, they left about one-fifth of the berries right on the vines. One man claimed that between a stone and a berry lying on the bog, the Darlington picker would take the stone every time. On newer bogs, mechanical harvesters couldn't get through the day without tearing up some turf. And being a complex combination of gears, teeth, chains, and engine parts, they broke down frequently. "The man who has three harvesting machines has two that work," the saying went. A large Wareham grower employed a worker to do nothing but repair picking machines during the harvest. But they did handle cranberries delicately. With a Darlington machine, it was possible for a frog to be scooped from the bog and onto a conveyor, then dumped into a box without the slightest injury.

Over the course of a decade after its introduction, the mechanical harvester slowly made incursions, overcoming grower uncertainty and the realm of the scoop until it was in use on the majority of cranberry bogs. It certainly wasn't inexpensive— it cost nearly a thousand dollars initially, the price dropping somewhat in later years—but it was less costly than scooping in the cranberry depression of the 1950s, when new technologies brought larger crops but demand remained stagnant. Yet economic considerations were not paramount to every cranberryman. Eastover Farms in the inland town of Rochester scooped their cranberries until 1964; the conservative streak was strong in the elderly grower who ran those bogs and died at eighty-seven.

Inevitably, the picking machine hurt workers who had come to rely on the cranberry harvest for wages each fall. Young Cape Verdeans were not among them. Culturally acclimatized to America, they sought and found better jobs at higher wages in the postwar economy. For men, this often meant employment in the building and construction trades. Cape Verdeans coming of age wanted nothing to do with the labor that had been their grandparents' and parents' lot on arriving in Massachusetts. It was their exodus from the cranberry bogs that helped to bring increasing numbers of picking machines to the harvest. For older Cape Verdeans, the story was very different. Often past the age of full employment, they had found scooping a useful income supplement, now gone. Doris Gomes of Marion saw the effect on a neighbor: "I remember the old lady down the street here, when they started with machines, she used to go watch them. And you'd see tears in her eyes."

For those tasks that still demanded vigorous young workers—ditching and sanding, and hauling berries off the bog—large growers turned to new immigrants just establishing themselves in America: Puerto Ricans. Some were migrant workers who wedged the cranberry harvest between picking Jersey tomatoes and gathering mushrooms in Pennsylvania. Others were flown up from Puerto Rico and spent the entire growing season working and living in alien New England. A few Puerto Rican laborers were recently settled in the area cities of New Bedford and Taunton. In America, when one immigrant group had risen from the low pay and long hours of farm work, there was always another to take its place.

Puerto Rican workers ditching.
Lindy Gifford photograph

A second technological change noted by Clarence Hall in his gentle criticism of Bay State cranberry growers as "slow to the new" was the advent of sprinkler irrigation. The technique of piping water onto cranberry bogs with electric or gasoline pumps and spreading it over the vines through rotating sprinkler heads made its appearance in the late 1930s. Sprinkler systems could serve the cranberry grower in many ways. They were a far more effective and penetrating means of watering bogs during hot spells and periods of drought than merely keeping water high in the ditches. Mixed with water, fungicides and insecticides could be applied quite efficiently through the system, putting the grower in contact with the poisons as briefly as possible. Finally, the sprinkler system was a marvelous and rapid deterrent to frost.

> ***The disadvantage of being the oldest producing area is that the status quo is sometimes foolishly accepted as desirable.... The real danger to Massachusetts is not that it will cease to be the biggest producing area, but that it will cease to be the best.***
>
> —Department of Agriculture
> *The Cranberry Industry in Massachusetts*

As a frost protector, the irrigation system operated on a principle known as the heat of fusion of water. When water was sprinkled on the bog and the air temperature was below freezing, a mantle of ice formed on the vines and buds or berries, depending on the season. Taken by itself, the encasing ice would drop the plants' temperature to below 32 degrees F and kill them. But endless sprinkling of the icy bog kept the air temperature above freezing. It requires a great deal of heat for water at 32 degrees F to become ice at 32 degrees F; the heat produced by water continuously freezing was absorbed by the surface of the iced plants, maintaining a temperature of 32 degrees F and keeping them from certain death. That this method of frost protection worked was remarkable in and of itself, but it also had decided advantages over flooding the bog.

Sprinkler irrigation used far less water to achieve the same results than flooding did, water that not every grower could pump back into his reservoir and conserve. More importantly, it saved the grower time and labor and allowed him to play fast and loose with the vagaries of nature. There was no way around starting the flood hours in advance of a frost, even if it never materialized. But defended by sprinklers, a grower could watch the mercury hourly, and if and only if a frost seemed imminent, start up his pump for immediate protection. The drawback to sprinkler systems in combat with frost was the frailty of the machinery. Once his bog was flooded, a cranberryman could collapse into bed and sleep peacefully. When his acreage was fitted out with sprinklers, he had to sit up as long as frost threatened or suffer the loss of his crop should the pump fail halfway through the night.

Cranberry growers on the Pacific Coast, in Washington and Oregon, quickly saw the advantages of sprinkler systems and began to install them in those early years. Massachusetts growers remained skeptical and lagged behind. In 1946 only 21 acres of bog were watered by sprinkler systems; by 1956, the number had grown to 350 acres, but this was still only a fraction of the 13,000 acres of producing cranberry bog in the state. At the close of that same year in Oregon and Washington, the majority of bogs had sprinklers.

What changed the minds of Massachusetts cranberrymen toward sprinkler irrigation was a natural catastrophe, a devastating frost that occurred on the evening of May 30, 1961, Memorial Day. The Cranberry Experiment Station issued a warning of a dangerous frost for that night, but the weather seemed to belie such a prediction. The day stayed overcast and breezy long after dark, usually the sign of a frostless night. But a fast-moving cold front changed all that. Between the hours of 10 and 11 P.M. it roared across Plymouth County and Cape Cod, in places slamming into the retreating warm air mass, causing thunderstorms. On its arrival, winds quickly died and the air temperature, which had remained in the forties all day, dropped well below freezing in thirty minutes' time. Many growers had taken the frost warning seriously despite the unlikely frost weather, and dutifully flooded their bogs. Yet many others chose to see what the evening brought before taking any action. When the evening brought a hard and killing frost and caught them unprepared, they could not flood quickly enough.

Some men lost an entire year's livelihood in that night; one-third of the state's crop, perhaps two hundred thousand barrels of cranberries, were taken by the weather. The May 30th frost of 1961 became legendary; it brought temperatures as low as nineteen degrees to some bogs and caused the vines to take on their ruddy winter coloration overnight. Properly, it was called a black frost, one in which not only the air temperatures but also the dewpoints were extremely low.

In less than one half-hour, Bay State cranberrymen learned what their counterparts in Washington and Oregon had known for years: without sprinkler systems there was no quick response to impending frost. Because the growers of the West had championed sprinkler irrigation early on, they were growing cranberries more efficiently, producing far more cranberries per acre. They watered the crop when too much cold brought frost and watered it again when too little rain made a drought. In 1964 Massachusetts bogs yielded fifty-seven barrels per acre; in Washington the average yield was sixty-seven barrels. Slowly, Massachusetts growers were led to water and made to irrigate. Five years after the great Memorial Day frost, one-third of the state's cranberry bogs were nurtured and protected by sprinkler irrigation.

Why were those who raised cranberries on Massachusetts bogs so wary of change, so suspicious of innovation? The agriculturists who wrote the 1968 bulletin *The Cranberry Industry in Massachusetts* for the state's Department of Agriculture knew the reason. They wrote:

> The disadvantage of being the oldest producing area is that the status quo is sometimes foolishly accepted as desirable. Marginal bogs slip into the category of money losers, as more efficient bogs are developed in newer producing areas. New growers are not slaves to old habits, and they generally adopt the latest and most efficient cultural and handling practices. The real danger to Massachusetts is not that it will cease to be the biggest producing area, but that it will cease to be the best.

The fabled conservatism that personified Massachusetts growers was not rooted in frugality or isolation so much as in tradition. Holding to old ways of growing and harvesting the cranberry was an expression of loyalty to earlier generations, of fidelity in doing things as one had been taught, as fathers and uncles and grandfathers had done. The burden of tradition grew heavier and heavier as the decades passed. The paradox of it all was that those who were appalled at the idea of canning cranberries in the early 1900s, who denied in the 1940s that a useful picking machine would ever be built, and who ignored sprinkler irrigation until 1961 were the grandsons and great-grandsons of innovators who exhausted every means in adapting the cranberry to cultivation during the mid-nineteenth century.

Hiller cranberry plant, Carver, Massachusetts.
Lindy Gifford photograph

Chapter 12

Modern Times

It was in 1959 that the world beyond Cape Cod and the other growing regions became acutely aware of the cranberry grower and his crop, and in that year, too, modern times first impinged upon the cranberry industry. Well into the Thanksgiving season, on November 9, secretary of Health, Education and Welfare, Arthur Fleming, revealed at a press conference that two shipments of West Coast cranberries—from Washington and Oregon—had been found to be contaminated with a weed killer called aminotriazole. In tests on animals, aminotriazole caused thyroid cancer, and for this reason, Fleming recommended that cranberries and cranberry sauce from these states be removed from supermarket shelves. Unfortunately, HEW could suggest no way for grocers or consumers to distinguish West Coast cranberries from those grown in Wisconsin, New Jersey, or Massachusetts. Overnight, the native cranberry lost its innocence and became a fruit of controversy and suspicion. The cranberry industry and each grower was sullied and thrust into the national consciousness. The cranberry crisis had begun. Even today, people who know not one other fact about the cranberry remember this event.

In response to the announcement, Ocean Spray's vice president, Ambrose Stevens, appeared on the *Today Show* with Dave Garroway to make a case for the purity and goodness of the vast majority of the crop. But public sentiment had turned against the cranberry. The evening news showed clerks removing fresh cranberries from supermarket shelves. Order cancellations flooded the growing region, causing packing plants to close down. Even the United States Army, an old friend of the cranberry, banned the fruit from Thanksgiving dinner and all other meals. In the first week of the cranberry crisis, sales of canned sauce were off 83 percent and fresh fruit sales down 71 percent. These figures improved by only a few percentage points the following week. It seemed as if the only remaining allies of the cranberry industry were the presidential candidates. Campaigning in Wisconsin, Senator John F. Kennedy drank cranberry juice and called for quick action, while Vice President Richard Nixon, also stumping for votes in the Badger State, downed four helpings of cranberry sauce. By November 12, eight lots of tainted cranberries totaling 84,000 pounds had been seized by the government. Four lots had originated in Wisconsin, two in Oregon, and one each in Washington and

Massachusetts. Only the crop from New Jersey's Pine Barrens remained free of any contamination. By Christmastime, sales of cranberries tested and certified pure picked up a little, but were still off by almost 50 percent compared to the previous December. And so the cranberry growers' most disastrous year petered out.

Cranberry growers had no one to blame but a few of their own number for the predicament. The herbicide aminotriazole was approved by the Department of Agriculture for use on cranberry bogs only after the harvest was complete, but a few men had applied it before the berries were picked, producing the telltale residue. The issue was not, as many growers thought, that less than one percent of the national crop had been contaminated, or that the link between aminotriazole and cancer in humans was less than firm. The real issue was that some farmers had risked endangering the national health by carelessly using a hazardous chemical.

For the innocent 99 percent of cranberrymen who grew pure fruit, the federal government provided some solace. The disaster relief was an indemnification program negotiated between the cranberry industry and the U.S. Department of Agriculture and instituted early in 1960. Under its terms, growers paid for the analysis of 1959's unsold crop with the understanding that the federal government would pay them for all fruit found to be uncontaminated. Lab technicians scrutinized upwards of half a million barrels of cranberries and the Department of Agriculture wrote checks totaling 8.5 million dollars to the nation's cranberry growers. The berries were destroyed, since no market remained for them and next season's crop was already on the vines, and the grower was reimbursed the cost of having produced them.

Here's a recipe that's proving very popular in New England this year. Mix large quantities of cranberries with gallons and gallons of gasoline or fuel oil. The concoction may sound uninviting for tomorrow's Thanksgiving feast, but cranberry growers apparently love it.

—*Wall Street Journal*

The immediate effects of the 1959 cancer scare were felt primarily by cranberry growers, who lost face and a year's profits, and the federal government, which spent money in order to combat a threat to public health and keep a group of farmers in business. But in a larger sense, the cranberry crisis was the dire tolling of a bell which has reverberated to this day. It presaged a now unending string of environmental crises all concerned with the presence of potentially hazardous chemicals in our water, air, and food. What seemed at the time an isolated incident was but our first awareness of man's recklessness with the things that science has allowed him to create.

In the years immediately following the aminotriazole scare, fate proved kinder to cranberrymen, but only by a little. Americans were again buying cranberries and sauce in the amounts they had before the scare, but this was not enough. While 1961 brought the great Memorial Day frost to Massachusetts, the growing region as a whole still produced a record crop of 800,000 barrels. Cranberry growers remained caught in the snare that had held them since the early 1950s: overproduction. New customers needed to be found for more and more berries, but until they were, cranberrymen would make little more on a barrel of fruit than the cost of production—ten dollars.

On the day before Thanksgiving in 1962, the *Wall Street Journal* reported a major step by the cranberry industry in solving the dilemma of too many berries. The article's title read: "Cranberry Growers' Thanksgiving Recipes Gather Crop, Oil Well." It began:

> Here's a recipe that's proving very popular in New England this year. Mix large quantities of cranberries with gallons and gallons of gasoline or fuel oil. The concoction may sound uninviting for tomorrow's Thanksgiving feast, but cranberry growers apparently love it. By thus destroying 12 percent of their 1962 crop, they've been able to push up the retail price for their berries as much as two cents a pound over 1961 levels even though this year's crop is a record.

The conflagration of cranberries alluded to by the *Journal* was the industry's end run around the overproduction problem. In 1962 a majority of the nation's cranberry growers voted for a federal marketing order which limited the amount of fruit each grower could sell. By "setting aside" 93,000 barrels of cranberries, as the industry euphemistically put it, or sending them up in smoke as the *Wall Street Journal* bluntly wrote, each grower got a little more money for the cranberries that were left.

The marketing order went against the grain for some men. It seemed wasteful and wrong to destroy perfectly good fruit, berries they had watered, protected against frost, and worried over. And once the *Wall Street Journal*'s story appeared, the marketing order became an embarrassment. By the early 1970s, a slowly increasing market combined with a decline in bog acreage meant that growers were producing only what could be sold, doing away with the controversial destruction of cranberries. This was good, because incinerating cranberries was no solution to overproduction. The solution to overproduction was greater consumption.

No one had to tell Ocean Spray Cranberries, Inc., now the country's dominant marketer, that it needed more customers. But the cooperative was just coming around to the realization that to find new customers it needed new products. Thirty years of experience now made it clear that no matter how much advertising appeared, sales of jellied and whole cranberry sauce would only increase so much each year. This was a fact, even if people stirred cranberry sauce into casseroles, dribbled it on top of ice cream, or ate it in some other novel way. Ocean Spray began looking to cranberry juice cocktail as the antidote to overproduction. The cocktail—a mixture of cranberry juice, water, and sweetener—had been sold since the early 1940s in a few regions but never marketed nationally. From the first, demand exceeded the supply, but Ocean Spray continued to produce the cocktail in small batches, believing that cranberry sauce would always be its most popular product. In 1963 Ocean Spray began selling cranberry juice cocktail nationwide, pushing it as a refreshing change from orange juice at breakfast. It bought air time on the *Arthur Godfrey Hour*—the host had invested in Massachusetts bog—and encouraged viewers to "Go Creative" with cranberry juice cocktail. Did Americans want to "Go Creative" with cranberry juice cocktail? Did they want a refreshing change from orange juice at breakfast? Time would tell.

With the 1960s, a problem from the recent past returned to haunt cranberry growers, and with them, every other human, mammal, fish, and bird in North America. The specter, again, was chemicals. But this time, the concern was not about their misuse, but about their very existence and the effect

Applying pesticides through an irrigation system.
Lindy Gifford photograph

that application of pesticides and herbicides had on living things. America's new fears about chemicals coalesced in a book published in 1962 and written by a woman named Rachel Carson. Her book was ominously titled *Silent Spring*. Carson, a biologist and acclaimed writer, focused on the pesticide DDT, which the army had used so successfully as a deterrent to typhus-carrying insects during World War II. Since then, it had been sprayed across the continent with near abandon by farmers and public health officials to kill a range of insect pests. Farmers could afford to use quantities of the pesticide, for besides being effective, it was very cheap.

Carson's warning was not only of DDT's toxicity, but of its persistence. The compound proved remarkably durable; it was stored in the fatty tissues of birds, fish, and mammals, and remained in the soil and water for years. While it was not clear that DDT actually produced cancer in humans, research showed that it killed migrating birds, diminished photosynthesis in aquatic plants, and inhibited the hatching of fish. Most notably, DDT was considered responsible for the demise of raptors. Ospreys, peregrine falcons, eagles, and other predatory birds compounded their DDT levels by feeding on contaminated prey; they then produced eggs with shells so fragile that most broke before hatching could occur. News of DDT's effects caused national alarm.

But the alarm felt by the nation's farmers—cranberry growers among them—was that they might lose a most effective combatant in the war against insects. Farmers and their spokesmen challenged the findings of researchers and asserted that as an agent in eradicating hunger and disease worldwide, DDT had done a great deal more good than harm. Rachel Carson was bitterly denounced for the environmental hysteria her book had created. But by 1971, Carson's distress signal about the effects of "elixirs of death" had worked its way to the heart of federal and state regulatory agencies. In that year, the use of DDT in Massachusetts was severely restricted and the newly created Environmental Protection Agency would soon ban other pesticides—Chlordane, Dieldrin, and Silvex—as potential human carcinogens. All the while, cranberry growers and cranberry researchers railed against the bannings as unnecessary and counterproductive, but to no avail.

Ironically, what replaced the banned chemicals are insecticides called organo-phosphates, compounds which break down more quickly after application and are meant to have a diminished effect on the environment, but are far more toxic to humans than DDT. Environmentalists retain their skepticism of cranberry growing, displeased by its historic exemption from the Clean Water Act and leery about waterborne fertilizer and pesticide residues presumably discharged into ponds, wetlands, and rivers. In the invention of new pesticides almost every victory is a pyrrhic one.

The cumulative effect of *Silent Spring* has been to permanently sensitize millions about the potential threat of chemicals to all ecosystems. Not since its publication have Americans watched a crop duster or mosquito control truck without suspicion. But even today, the name Rachel Carson raises a red flag in the minds of cranberry growers, who do not believe or do not want to believe, that chemicals they found so useful could also do great harm.

Invention proceeds indifferent to questions of public policy, so that the times that brought the social dilemma of pesticides also ushered in the last great technological innovation to yet grace Massachusetts cranberry bogs: water harvesting. But while Cape Cod growers were just beginning to accept it as a boon to their livelihood, their counterparts in Washington, Oregon, and Wisconsin were perfecting the technology they had embraced a decade earlier. Even in New Jersey's Pine Barrens, where old ways also died hard, growers experimented with water harvesting a full five years before Massachusetts men seriously considered the new method.

At the heart of water harvesting was the simple fact that the cranberry, with its interior air sack, floats in water. Growers had made use of this characteristic for years. Wisconsin, with its abundant water supplies, had developed an efficient scooping technique on flooded bogs. With the vines immersed in water, the berries floated upward, free of the jungle of vines which hid them on a dry bog. Pickers in hip boots then bent to scoop them "on the flood." The wet berries were dried on long racks at the bog edge. In the other growing regions, cranberrymen used water harvesting to collect that portion of the crop left on the bog after the harvest, if the market warranted it. Again the bog was flooded and the berries previously dropped or knocked off the vines were corralled with floating booms and collected with a wooden strainer. The work was called "taking floaters."

The first great advantage of water harvesting was that the grower picked nearly every cranberry he'd grown, thanks both to water and a novel machine. Developed in the West, the harvester was a slender steel frame with wheels and a gasoline engine to which a metal reel had been attached at the front. It came first in walking, and then riding, models. In operation, its motion resembled nothing so much as a running eggbeater held sideways. Churning the water, the reel vibrated some berries and pummeled others off the vines and up to the surface. In the harvester's wake, men in waders herded the floating crop together, encircled it with booms, and led it to the shore. This was work that a prevailing wind did just as nicely. At the shore, a motorized conveyor or vacuum hose sent the berries up and into a sizable truck which chauffeured them on to a processing plant.

The savings in time and money that water picking brought to Pacific Coast, Wisconsin, and Jersey bogs exceeded the savings growers realized in switching from the scoop harvest to machine harvesting. A water-picking gang was no larger in number than a dry-picking crew, but it could harvest considerably more acreage in an equal amount of time. And while the dry-picking machines left a quarter of the crop on the vines, water harvesting captured virtually every cranberry. In 1964 Massachusetts growers got fifty-seven hundred-pound barrels to the acre. Their counterparts in Wisconsin harvested ninety-six barrels per acre, and those in Washington and Oregon sixty-seven and sixty-six barrels, respectively. The difference was water harvesting.

Even in the long years of cranberry surpluses and marketing orders, water harvesting made sense. It increased the yield of cranberries so greatly that growers could now afford to abandon marginal bogs that produced little. And an early familiarity with this technology prepared cranberrymen for the day when demand for cranberries would outstrip supply, and each grower would take to market every cranberry he could get his hands on. From the perspective of the mid-1960s, that day seemed a ways off, but not just a dream. Demand for the cranberry had increased slowly but steadily, and Ocean Spray was now introducing mixed juice drinks called Cranapple and Crangrape, which sparked curiosity in the marketplace.

Massachusetts cranberrymen had their reasons for coming around so slowly to water harvesting. Some of the reasons were practical while others were philosophical or, perhaps, emotional. First, early autumn rainfall on the Cape was always light, and some water that growers had managed to store inevitably went to protect ripening berries on cold nights. This meant that water for harvesting was not abundant. Then there was the problem of coverage. Many Massachusetts bogs were ancient and badly out of grade. At one end the water might be so deep that the machines could not reach the berries, while at the other, the bog sat high and dry. But these were problems that most growers—with a little planning and work—could overcome, and eventually did.

The greater stumbling block was their concern for the stability of the declining fresh fruit market. Water harvesting was a technology designed specifically to gather fruit for processing, the means by which the great majority of the crop was now prepared for market. The work of the reels softened the berries slightly, a matter of no consequence for fruit soon to be made into sauce or juice. But water harvesting meant that a grower's crop could only be sold for processing, while dry harvesting allowed him the opportunity to serve either outlet. The first and traditional market for the cranberry was as fresh fruit, and many men were loath to leave that

Water harvesting, Maple Springs Bog, Wareham, Massachusetts.
Lindy Gifford photograph

Booming in cranberries, Piney Wood Bogs, Carver, Massachusetts.
Lindy Gifford photograph

market behind by changing their harvesting method. It was as if in keeping their old picking machines, they dreamed of a resurgence in the fresh cranberry trade, a day when store counters would again brim with boxes and bags of ripe fruit. This dream was not to be, and during the late 1960s and 1970s Massachusetts cranberry growers slowly, perhaps painfully, made the transition to water harvesting. In order to make their bogs more productive, compete with other cranberry growing states, and respond to the market, it was a change that was inevitable.

Water picking has had one quite unanticipated effect on the cranberry industry—it has brought tourists to the bogs. Every autumn they pull to the side of the road and stare at an improbable sight: weird machines moving across an inland sea churning up streams of crimson berries. Farmers in fishing waders slog around, herding the fruit together into great red rafts. And in the air, a complex smell of muck and water, of plants and fruit and gasoline. It is cold, wet work that people watch from the warmth of their cars. Harvesters spend the day encased in clammy rubber, and sometimes in discomfort if they step into a ditch and water fills their waders. On a raw, wet day, the work of water harvesting is harsh.

We went to Grand Canyon and we didn't see as bright a colors or as many of them as we're going to see in our own harvest.... Our bogs ain't one of the Seven Wonders of the World, but when it comes to color, you can't beat the cranberry harvest. It is spectacular.

—Larry Cole

On Cape Cod, where the foliage is masked by so many pine trees, the bog is also home to fall's most striking colors. For Larry Cole, the late Carver cranberryman and raconteur, this was emphatically true. "We went to Grand Canyon and we didn't see as bright a colors or as many of them as we're going to see in our own harvest. Those colors are subdued by comparison. Our bogs ain't one of the Seven Wonders of the World, but when it comes to color, you can't beat the cranberry harvest. It is spectacular."

Perhaps the one thing more spectacular than the bog at harvest time was the return that cranberries brought to growers for almost a generation, from the late 1970s forward. Since the mid-1950s, growing cranberries had been a marginal investment, the price moving up ploddingly. But in 1978, the cranberry market made a definitive move. Demand finally outstripped supply and the price per barrel shot from thirteen dollars up to nearly twenty dollars. All of a sudden there was money in cranberries.

It was the growing popularity of Ocean Spray's juice drinks that finally made the cranberry a favored fruit. Sold nationally for a decade or more, the beverages were now in sync with the national taste. Wary of the unnatural, Americans veered away from soda pop and toward fruit juices. It was not just a faddish taste; the demand for juice drinks of every fruit imaginable continues unabated. Ocean Spray has ridden this wave with great skill, blending the cranberry with apples, mangos, wild blueberries, tangerines, and grapes to produce a variety of popular juices, while also marketing juices unallied with the cranberry— grapefruit, for example. With a product line greatly expanded beyond the corporate name Ocean Spray Cranberries, Inc., the company commands 65 percent of the continent's juice and juice drink market. Growers thank their stars that the cooperative shifted its attention to marketing drinks, for this made cranberry growing profitable agriculture. The spirit of Marcus Urann prevailed.

As luck would have it, the rise in the price of cranberries dovetailed perfectly with a great increase in yield from the bogs. Sprinkler systems providing both irrigation and frost protection combined with the switch to water harvesting have made the difference. In 1967 the average number of barrels harvested per acre was 50; it grew to 150. Good fortune brought great change to cranberry growing. Once there had been many bogs for sale and few buyers; now the reverse became true and a single acre of producing cranberry bog commanded $30,000. New faces and new money came to the industry and old names returned. The largest bog properties on the market were purchased, as they had been at the turn of the century, by investors who left growing to a local cranberryman. Young couples, drawn by the outdoor work and high profits, bought up small bogs. Sons and daughters of growing families who left when the industry was down returned to learn the trade.

Processing cranberry juice, Ocean Spray Cranberries, Inc.
Lindy Gifford photograph

New investment worked to change the appearance and, perhaps, the character of the region itself. Those with space for new bogs built them—the land cut over, the stumps grubbed up—and reclaimed plots abandoned in the long cranberry depression. At bogside, new buildings rose. Weathered storage sheds and screenhouses clad in native pine or cedar gave way to metal structures in blue and green. The precious quality of this landscape, of historic bogs and buildings matched, began to fade. Among the growers there was unaccustomed consumerism on and off the bogs. Shiny tractors mowed the uplands while tiny excavators inched along the ditches scooping up weeds and muck. A grower's pickup might be cherry red with balloon tires and a roll bar. The older growing families, remembering the cranberry's past advances and declines, approached this great cranberry revival with caution. Their homes, cars, and lives seemed much the same as before, save for longer winter vacations and help from financial planners. Among younger growers, the prosperity became tradition, and consumption more conspicuous. Some drove Mercedes or built striking contemporary houses; still others skied in Switzerland or body-surfed in Maui.

Would the good times last? The good times would be sorely tested. Wisconsin growers, with a long cranberry tradition of their own, responded to the market with a furious and sustained expansion of "marshes"—not bogs. For decades, Wisconsin had trailed Massachusetts in acreage and production, but its vast freshwater wetlands made it the cranberry industry's new frontier. During the mid-1990s, nearly 1,000 acres of marsh was built yearly, and Wisconsin surpassed the Bay State in cranberry acreage. The fever traveled Down East, too. Maine was known for wild blueberry culture, a trade even more antique and hidebound than the cranberry business. But blueberry growers and others were intrigued by the dramatic returns and experimented with small bogs. When all of the new cranberry real estate—representing New Jersey, Washington, and Oregon, as well—bore fruit, production once again exceeded consumption. A barrel of cranberries fetched sixty-six dollars in 1996, but only seventeen dollars in 2000.

Diversification is one message that comes from the shock of economic cycles, and some in the cranberry industry have heard it. Organic cranberries are coming to the fore, and a few growers, like Cranberry Hill in Plymouth, Massachusetts, have been producing this way for years. The industry's early cultural practices, spring flooding, and sanding to bury insect eggs, are tried and true organic methods. Equal Exchange, the fair-trade coffee cooperative, also markets organic dried and sweetened cranberries—sourced from both Wisconsin and Buzzards Bay, Massachusetts. Dried cranberries themselves are a recent growth sector in the trade. Decas Cranberry Products, Inc., founded in the 1930s by a family of independent and relentless innovators, has fashioned itself into "The Cranberry Ingredients Experts" and sells million of pounds of dried fruit that brighten and fortify General Mills cereals and other foods. The company is also perfecting nurtaceuticals derived from cranberry, advancing its value as an antioxidant. A new cranberry product with unparalleled authenticity is the Tanka Bar. The delicious treat is modern pemmican, combining cured buffalo with dried Oregon and Wisconsin cranberries and made by Oglala Lakotas on the Pine Ridge Indian Reservation in South Dakota.

One approach to income diversification underway in Massachusetts's cranberry industry is bold and controversial. A. D. Makepeace Company, the region's oldest grower, is now a real estate developer. The lesson it took from the most recent cranberry boom and bust was the need to establish additional profitable businesses to satisfy shareholders. With 12,000 acres in Plymouth and Barnstable Counties, Makepeace is the largest private landowner in eastern Massachusetts. It grows cranberries on one-sixth of this acreage, managing the white-pine woodlands for timber on another 4,000 acres. The model of a natural resource business turning to land development is not unknown. Plum Creek Timber Co., Inc., of Seattle is the nation's largest private landowner and routinely identifies its properties with high scenic or resource value for residential development or sale to conservation interests. To date, Makepeace has developed a half-dozen upper income subdivisions tucked into the bog, pond, and pinewoods landscape—attractive, but isolated developments. Its real estate advertisements in the *Boston Sunday Globe* read, "We've been getting your backyard ready for 150 years," but this is not so. Makepeace has long been a good neighbor in Plymouth County towns, permitting access to its lovely, distinctive lands for walking or fishing. But until recently, the pine uplands abutting cranberry bogs were considered a precious buffer and encroaching residences an anathema. The acres of woodland separated homeowners from the off-chance of pesticide drift and the late-night roar of a pump engine. They also protected the valuable fresh

Abandoned bog ringed by homes, Hanson, Massachusetts.
Lindy Gifford photograph

water in ponds and wetlands that growers manage for irrigating and flooding. To operate a business that derives income from agricultural and residential development simultaneously is tricky.

The greatest challenge for Massachusetts and for Makepeace lies ahead. The company has proposed phased development of 1,175 homes with accompanying shops, offices, village green, and walking trails in Plymouth, Massachusetts—what planners refer to as a "new town." Population in southeastern Massachusetts, the fastest-growing part of the state, is projected to increase by 200,000 in a generation and the Makepeace project may be its coda. The context for the development is this: over the past thirty years the region has lost a third of its forest and agricultural lands.

Recognition of the cranberry landscape evolved differently in another growing region. It was John McPhee's evocative *The Pine Barrens,* published in 1967, that drew attention to the quiet, rural land of south central New Jersey and a distinctive culture, flora, and fauna. From this murmur of appreciation grew a campaign to protect its resources and plan their use. The New Jersey Pinelands Commission now oversees these assets, which span a million acres of land atop an aquifer of seventeen trillion gallons. Public lands and small towns are interspersed; blueberry and cranberry operations mix with cedar swamp, rare native orchids, and sandy tracts of pine and oak. This, the greatest stretch of open space in the mid-Atlantic, has been designated a UNESCO biosphere reserve.

New eyes are watching southeastern Massachusetts, as well, and seeing the landscape in a different light. The Nature Conservancy has termed the region's own 20,000 acres of pine barrens the Plymouth Pinelands and drawn attention to the vast and invaluable aquifer sitting beneath it. The rare Atlantic white cedar swamps, coastal plain ponds, and river otters of the New Jersey Pine Barrens are indigenous here, too, and warrant preservation.

Both the physical and social environment in which cranberry growers carry on a more than century-old agricultural tradition is evolving rapidly. Modern times are drawing cranberry growing, development, and conservation into the same tight circle. The question at hand is whether a balance can be found between continuity and change.

Bog landscape, Plymouth, Massachusetts.
Lindy Gifford photograph

The Cranberry Room, circa 1983, Middleborough Public Library, Massachusetts.
Lindy Gifford photograph

Bibliography

* Note: All interviews included in the bibliography are available for use at the institutions listed. In the Collections of the Northeast Archives of Folklore and Oral History, Maine Folklife Center, they can be located by accession number, while in the W. B. Nickerson Archives, Wilkins Library, at Cape Cod Community College the tapes are filed by the surname of the person interviewed.

Wild Harvests

Appelbaum, Diana Karter. *Thanksgiving: An American Holiday, An American History*. New York: Facts On File, 1984.

Baker, James W. "Recreating the First Thanksgiving Dinner," Research Library, Plimoth Plantation, Plymouth, MA.

Bradford, William. *Of Plimoth Plantation*. Edited by S. E. Morison. New York: Knopf, 1952.

[By a Lady.] *A New System of Domestic Cookery*. Boston: William Andrews, 1807.

Chamberlain, N. H. *An Autobiography of a New England Farm House: A Romance of the Cape Cod Lands*. Boston: Cupples & Hurd, 1888.

Child, Mrs. *The American Frugal Housewife*. Boston: Carter & Hendee, 1832.

Collections of the Massachusetts Historical Society. Vol. III, Second Series. Boston: John Eliot, 1815.

"Cranberry Day," *Boston Transcript*, 26 September 1831.

Deetz, James, and Jay Anderson. "The Ethnogastronomy of Thanksgiving," *Saturday Review of Science*, November 25, 1972.

Dunn, Richard S. *Sugar and Slaves: The Rise of the Planter Class in the British West Indies, 1624–1713*. University of North Carolina Press: Chapel Hill, 1972.

Forbush, Edward H. *Birds of Massachusetts and Other New England States*. Part I. Issued by the Legislature, 1925.

Hartlib, Samuel. *His Legacy of Husbandry*. England, 1655.

Huntington, Gale. *An Introduction to Martha's Vineyard*. Edgartown, MA: Dukes County Historical Society, 1974.

Josselyn, John. *New England's Rarieties Discovered*. London, 1672; reprint Boston: Veazie, 1865.

Markham, Gervase. *Country Contentments or The English Huswife*. London, 1614.

Murrell, John. *A New Booke of Cookerie*. London, 1615.

New Columbia Encyclopedia. New York: Columbia University Press, 1975.

Oxford English Dictionary. Volume II, Oxford, England: Clarendon Press, 1961.

Pinney, H. *Family Receipts*. Philadelphia, PA, 1848.

Records of the Colony of New Plymouth in New England. Court Orders Vol. IV, 1641–1651. Edited by Nathaniel B. Shurtleff. Boston: William White, 1855.

Reynard, Elizabeth. *The Narrow Land; Folk Chronicles of Old Cape Cod*. Chatham, MA: The Chatham Historical Society, Inc., 1978.

Russell, Howard S. *Indian New England Before the Mayflower*. Hanover, NH: University Press of New England, 1980.

Simmons, Amelia. *American Cookery*. Hartford, CT: Hudson & Goodwin, 1796.

Simmons, William S. *Spirit of the New England Tribes: Indian History and Folklore, 1620–1984*. Hanover, NH: University Press of New England, 1986.

"The Cranberry Is for All Seasons," *Los Angeles Times*, 28 November 1985.

Thoreau, Henry D. *The Journal of Henry D. Thoreau*. Edited by Bradford Torrey and Francis H. Allen. Fourteen volumes, bound as two. New York: Dover Publications, Inc., 1962.

———. *The Natural History Essays*. Edited by Robert Sattelmeyer. Salt Lake City, UT: Peregrine Smith, Inc., 1980.

Tudor, William. *William Tudor: Miscellanies.* Boston, 1821.

Vanderhoop, Leonard F. Gay Head, Massachusetts. Interview, Fall 1983. Accession 1770, Collections of the Northeast Archives of Folklore and Oral History, Maine Folklife Center, University of Maine, Orono, ME. Also W. B. Nickerson Archives of Cape Cod History, Wilkens Library, Cape Cod Community College, Barnstable, MA.*

Wilkins, Ruth C. *Carlisle: Its History and Heritage.* Carlisle: Carlisle Historical Society, 1976.

Williams, Roger. *A Key Into the Language of America.* London: Gregory Dexter, 1643; reprint Providence, RI, and Providence Plantations Tercentenary Committee, Inc., 1936.

Wood, William. *New England's Prospect.* London, 1635; reprint Boston: Prince Society, 1865.

Cranberry Fever

Albion, Robert, William Baker, and Benjamin Labaree. *New England and the Sea.* Mystic, CT: Mystic Seaport Museum, Inc., 1972.

Boston Traveler. 24 July 1888.

Byers., Douglas S., ed. *The Prehistory of the Tehuacan Valley. Volume One. Environment and Subsistence.* Austin: University of Texas Press, 1967.

Casarjian, Conrad K. "A Geographical Analysis of Cranberry Bog Distribution in Massachusetts." Master's Thesis, University of
Colorado, 1967.

Chamberlain, Barbara Blau. *These Fragile Outposts—A Geological Look at Cape Cod, Martha's Vineyard, and Nantucket.* New York: Natural History Press, 1964.

Chandler, F. B. and Irving Demoranville. *Cranberry Varieties of North America.* Bulletin 513. Experiment Station, College of Agriculture, University of Massachusetts, 1958.

Collections of the Massachusetts Historical Society. Volume VIII. Boston: Munroe & Francis, 1802.

Cranberries. Edited by Clarence J. Hall. Wareham, MA: Wareham Courier, 1936–1966.

"Cranberries." *The Mirror.* Nantucket, MA, 1859.

Eastwood, Rev. Benjamin. *The Cranberry and Its Culture.* New York: C. M. Saxton Co., 1859.

Eleventh Annual Report of the Secretary of the Massachusetts Board of Agriculture for 1863. Boston: Wright & Potter, 1864.

First Annual Report of the Secretary of the Massachusetts Board of Agriculture for 1853. Boston: William White, 1854.

First Report on the Agriculture of Massachusetts, 1837. Boston: Dutton & Wentworth, 1838.

Holmes, O. M. "Cape Cod Cranberry Methods." In *American Cranberry Growers Association Annual Report 1883.* New Jersey: 1883.

Jorgensen, Neil. *A Sierra Club Naturalist's Guide: Southern New England.* San Francisco: Sierra Club Books, 1978.

Mason, Carol Young. "The Geography of the Cranberry Industry in Southeastern Massachusetts." Master's Thesis, Clark University, 1925.

Report of the Joint Special Committee Upon the Subject of the Flowage of Meadows on Concord and Sudbury Rivers. Boston: William White,1860.

Russell, Howard S. *A Long, Deep Furrow: Three Centuries of Farming in New England.* Hanover, NH: University Press of New England, 1982.

Third Annual Report of the Secretary of the Massachusetts Board of Agriculture for 1855. Boston: William White, 1856.

Thirteenth Annual Report of the Secretary of the Massachusetts Board of Agriculture for 1865. Boston: Wright & Potter, 1856.

Thoreau, Henry D. *The Journal of Henry D. Thoreau.* Edited by Bradford Torrey and Francis H. Allen. Fourteen volumes, bound as two. New York: Dover Publications, Inc., 1962.

Transactions of the Agricultural Societies in the State of Massachusetts for the Year 1849. Boston: Dutton & Wentworth, 1850.

White, Joseph J. *Cranberry Culture.* New York: Orange, Judd, 1912.

A Way of Life

Baker, Archelaus. Account Book, ca. 1890–1910. Archives, Cape Cod Community College, Barnstable, Massachusetts.

Bradley, James W., ed. *Historic and Archaeological Resources of Southeast Massachusetts:* Massachusetts Historical Commission, 1982.

Burgess, Nathan B. Letter to Professor C. H. Fernald, 16 September 1891. University of Massachusetts Cranberry Experiment Station, East Wareham, MA.

Burrows, Fredrika A. *Cannonballs and Cranberries.* Taunton, MA: Wm. S. Sullwold, 1976.

Carleton, Amy, Augusta Carleton Jillson, and Annie Carleton Lloyd. East Sandwich, Massachusetts. Interview, Winter 1983. Accession 1647.

Chamberlain, Barbara Blau. *These Fragile Outposts—A Geological Look at Cape Cod, Martha's Vineyard, and Nantucket.* New York: Natural History Press, 1964.

Cranberries. Edited by Clarence J. Hall. Wareham, MA: Wareham Courier, 1936–1966.

Cranberry Gathering on Cape Cod. Sandwich, MA: Seaside Press, 1879.

Crowell, William. Cranberry Pickers. Patent No. 157,158. U.S. Patent Office, Washington, D.C.

Crowell, William E. Dennis, Massachusetts. Interview, Spring 1983. Accession 1659.

Deyo, Simeon L., ed. *History of Barnstable County, Massachusetts.* New York: H. W. Blake & Co., 1890.

Griffith, Henry S. *History of the Town of Carver, Massachusetts, 1637–1910.* New Bedford, MA: E. Anthony & Sons, 1913.

Holmes, O. M. "Cape Cod Cranberry Methods." In *American Cranberry Growers Association Annual Report 1883.* New Jersey: 1883.

Jesus, Antone. Onset, Massachusetts. Interview, Winter 1983. Accession 1652.

Makepeace, Maurice, B. Wareham, Massachusetts. Interview, Spring 1983. Accession 1665.

New Yorker. "Notes & Commentary." 24 May 1958.

Peterson, Byron S., Chester E. Cross, and Nathaniel Tilden. *The Cranberry Industry in Massachusetts.* Bulletin 201. Massachusetts Department of Agriculture, 1968.

Penti, Marsha. "A Narrative History of the Town of Carver." Typescript prepared for Carver Historical Commission, 1980.

Redfield & Son. Postcards to cranberry growers, 29 December 1888 and 20 November 1895. Harwich Historical Society, Harwich, MA.

Ryder, Malcolm E. *History of the Cranberry Industry in the Cotuit Area.* Historical Society of Santuit and Cotuit, 1966.

Ryder, Malcolm, and Katherine Ryder. Cotuit, Masschusetts. Interview, Winter 1983. Accession 1650.

Seven Villages of Barnstable, The. Town of Barnstable, 1976.

Simonds, Sue Carolyn. "The Microgeography of Cranberry Farming in Plymouth County, Massachusetts" Master's Thesis, Clark University, 1971.

Simpson, Mary Peterson. Harwich, Massachusetts. Interview, Fall, 1982. Accession 1645.

Stillman, Ellen. Hanson, Massachusetts. Interview, Spring 1983. Accession 1664.

Tenth Annual Convention of the New Jersey Cranberry Growers Association. 1882.

Trayser, Donald G. *Barnstable: Three Centuries of a Cape Cod Town.* Hyannis, MA: Goss, 1939.

Webb, James. *Cape Cod Cranberries.* New York: O. Judd Co., 1886.

Tools of the Trade

Bailey, Eunice, and Jennie (Bailey) Shaw. South Carver, Massachusetts. Interview, Spring 1983. Accession 1767.

Buzby, J. Cranberry Separator. Patent No. 173,583. U.S. Patent Office, Washington, D.C.

Cranberries. Edited by Clarence J. Hall. Wareham, MA: Wareham Courier, 1936–1966.

Ellis, Ernest Clifton. *Reminiscences of Ellisville.* Plymouth, MA: Memorial Press, 1973.

Equipment for the Cranberry Grower. H. R. Bailey Company, South Carver, MA.

Glassie, Henry. *Pattern in the Material Folk Culture of the Eastern United States.* Philadelphia: University of Pennsylvania Press, 1968.

Hall, L. & Z., and W. Crowell. Cranberry Picker. Patent No. 199,703. U.S. Patent Office, Washington, D.C.

Hall, W. Cranberry Gatherer. Patent No. 81,897. U.S. Patent Office, Washington, D.C.

Hayden, L. A. Cranberry Separator. Patent No. 661,801. U.S. Patent Office, Washington, D.C.

Hayden Cranberry Separator Mfg. Co., The. Plymouth, MA: Memorial Press, 1925.

Holmes, O. M. "Cape Cod Cranberry Methods." In *American Cranberry Growers Association Annual Report 1883.* New Jersey, 1883.

Howes, Ernest D. Wareham, Massachusetts. Interview, Winter 1983. Accession 1649.

Lumbert, D. Cranberry Gatherer. Patent No. 289,846. U.S. Patent Office, Washington, D.C.

———. Cranberry Gatherer. Patent No. 654,013. U.S. Patent Office,Washington, D.C.

Russell, Howard S. *A Long, Deep Furrow: Three Centuries of Farming in New England.* Hanover, NH: University Press of New England, 1982.

Ryder, Malcolm, and Katherine Ryder. Cotuit, Massachusetts. Interview, Winter 1983. Accession 1650.

Sturdy, Joseph E. Cranberry Gatherer. Patent No. 489,239. U.S. Patent Office, Washington, D.C.

Thatcher, Charles. Cranberry Gatherer. Patent No. 48,136. U.S. Patent Office, Washington, D.C.

Washburn, J. S. Cranberry Picker. Patent No. 585,455. U.S. Patent Office, Washington, D.C.

Waters, W. B. Cranberry Gatherer. Patent No. 649,377. U.S. Patent Office, Washington, D.C.

Webb, James. *Cape Cod Cranberries.* New York, O. Judd Co., 1886.

Out of the Islands

Almeida, Raymond Anthony, ed. *Cape Verdeans in America: Our Story.* Boston: TCHUBA, The American Committee for Cape Verde, Inc., 1978.

Almeida, Raymond A., and Patricia Nyhan. *Cape Verde and Its People: A Short History.* Boston: The American Committee for Cape Verde, Inc., 1976.

"Cape Cod Africans," *Falmouth Enterprise,* Falmouth, MA, 25 August 1944.

Erickson, Charlotte. *American Industry and the European Immigrant, 1860–1885.* Cambridge: Harvard University Press, 1957.

Geller, L. D., ed. *They Knew They Were Pilgrims: Essays in Plymouth History.* New York: Poseidon Books, 1971.

Leavitt, John F. *Wake of the Coasters.* Middletown, MA: Wesleyan University Press for the Marine Historical Association, Inc., 1974.

Nunes, Maria Luisa, ed. *A Portuguese Colonial in America: Belmira Nunes Lopes. The Autobiography of a Cape Verdean American.* Pittsburgh, PA, Latin American Literary Review Press, 1982.

Peterson, Byron S., Chester E. Cross, and Nathaniel Tilden. *The Cranberry Industry in Massachusetts.* Bulletin 201. Massachusetts Department of Agriculture, 1968.

Pina, Vincent, and Beatrice Pina. Marion, Massachusetts. Interview, Fall 1983. Accession 1769.

"Sovoia Arrives," *The New Bedford Evening Standard,* New Bedford, MA, 5 October 1914.

Tyack, Bruce David. "Cape Verde Immigration to the United States." Senior Thesis, Harvard College, 1952.

U.S. Congress. Reports of the Immigration Commission. *Immigrants in Industries* (in 25 parts). *Part 24: Recent Immigrants in Agriculture* (in 2 volumes; Vol. II). 61st Congress, 2nd Sess., Senate Document No. 633. Washington, D. C.: GPO, 1911.

Living a Postcard

Bailey, Eunice, and Jennie (Bailey) Shaw. South Carver, Massachusetts. Interview, Spring 1983. Accession 1767.

Beaton, Gilbert B. Buzzards Bay, Massachusetts. Interview, Summer 1983. Accession 1768.

"Cape Pickers Union Formed," *The Standard Times,* New Bedford, MA, 6 June 1933.

Child Labor Facts. New York: National Child Labor Committee, 1924.

"Cranberry Pickers May Have Union." *Wareham Courier,* Wareham, MA, 1 September 1933.

"Crisis in Cranberry Strike Approaching as Week of Turbulence Ends." *Wareham Courier,* 15 September 1933.

Garside, Edward B. *Cranberry Red.* Boston: Little, Brown, 1938.

———. Plymouth, Massachusetts. Interview, Spring 1983. Accession 1661.

Gomes, Doris. Marion, Massachusetts. Interview, Winter 1983. Accession 1648.

Howes, Ernest D. Wareham, Massachusetts. Interview, Winter 1983. Accession 1649.

New York Times, 14 September 1933.

Pina, Vincent, and Beatrice Pina. Marion, Massachusetts. Interview, Fall 1983. Accession 1769.

Spinner: People and Culture in Southeastern Massachusetts. New Bedford, MA: Reynolds-Dewalt Printing, Inc., 1984.

"Strikers Form Picket Lines." *The Standard Times,* New Bedford, MA, 11 September 1933.

U.S. Congress. Reports of the Immigration Commission. *Immigrants in Industries* (in 25 parts). *Part 24, Recent Immigrants in Agriculture* (in 2 volumes; Vol.II). 61st Congress, 2nd Session, Senate Document No. 633. Washington, D. C.: GPO, 1911.

Wareham Courier, Wareham, MA, 8 September 1933.

———. 22 September 1933.

———. 29 September 1933.

———. 6 October 1933.

Tony Jesus

Jesus, Antone. Onset, Massachusetts. Interview, Winter 1983. Accession 1652.

From Finland

Bailey, Eunice, and Jennie (Bailey) Shaw. South Carver, Massachusetts. Interview, Spring 1983. Accession 1767.

Cranberries. Edited by Clarence J. Hall. Wareham, MA: Wareham Courier, 1936–1966.

Harju, Wilho, and Lillian Harju. Carver, Massachusetts. Interview, Spring 1983. Accession 1662.

The Raw and the Cooked

Cole, Lawrence, and Ruth Cole, North Caver, Massachusetts. Interview, Spring 1983. Accession 1660.

Hobson, Asher, and J. Burton Chaney. *Sales Methods and Policies of a Growers' National Marketing Agency: A Study of the Organization and Achievements of Twenty-Six Years of Cooperative Marketing by Part of the Cranberry Growers of the United States.* Bulletin 1109. U.S. Department of Agriculture, 1923.

Howes, Ernest D. Wareham, Massachusetts. Interview, Winter 1983. Accession 1649.

"Huge Surplus of Cranberries Deplored." *Middleboro Gazette,* Middleboro, MA, 23 April 1948.

Knapp, Joseph G., and Associates. *Great American Cooperators: Biographical Sketches of 101 Major Pioneers in Cooperative Development.* Washington: American Institute of Cooperation, 1967.

Middleboro Gazette, Middleboro, MA, 11 February 1949.

Monthly Review. Federal Reserve Bank of Boston. September 1953.

"Ocean Spray: The Early Years, 1930–1945." *Harvest,* Volume II, No. 2, Spring 1980.

Pettier, George L. *A History of the Cranberry Industry in Wisconsin.* Detroit, MI: Harlo Press, 1970.

Peterson, Byron S., Chester E. Cross, and Nathaniel Tilden. *The Cranberry Industry in Massachusetts.* Bulletin 201. Massachusetts Department of Agriculture, 1968.

Quarles, John. "Ocean Spray, 1930–1960." An address at the annual meeting of Ocean Spray Cranberries, Inc. August 17, 1960.

Report of Survey: Cranberry Canners, Inc. Booz, Allen & Hamilton. 1945.

"Rules for Branding." New England Cranberry Sales Co.

Russell, Howard S. *A Long, Deep Furrow: Three Centuries of Farming in New England.* Hanover, NH: University Press of New England, 1982.

Stillman, Ellen. Hanson, Massachusetts. Interview, Spring 1983. Accession 1664.

The Ocean Spray Story. Ocean Spray Cranberries, Inc., Plymouth, MA.

Urann, Marcus L. "Cranberry Canners, Inc. On the Air." Weekly broadcasts delivered by Marcus L. Urann, president, Cranberry Canners, Inc., to Massachusetts cranberry growers between 20 September and 27 December 1944 on stations WNBN, New Bedford, and WOCB, Cape Cod.

Wareham Courier, Wareham, MA, 22 August 1946.

The Grower's Almanac

Annual Report, 1943. Cape Cod Cranberry Growers Association.

Brackett, Philip. Cotuit, Massachusetts. Interview, Spring 1983. Accession 1666.

Cole, Lawrence, and Ruth Cole. North Carver, Massachusetts. Interview, Spring 1983. Accession 1660.

Cranberries. Edited by Clarence J. Hall. Wareham, MA: Wareham Courier, 1936–1966.

Crowell, William E. Dennis, Massachusetts. Interview, Spring 1983. Accession 1659.

Franklin, Henry J. *Cranberry Growing in Massachusetts.* Extension Leaflet No. 72. Massachusetts Agricultural College, Amherst, 1923.

Howes, Ernest D. Wareham, Massachusetts. Interview, Winter 1983. Accession 1649.

Kiernan, Richard. Wareham, Massachusetts. Interview, Winter 1983. Accession 1646.

Lazell, James D., Jr. *This Broken Archipelago.* New York: Quadrangle/The New York Times Book Co., 1976.

Rounsville, W. Marland. Nantucket, Massachusetts. Interview, Spring 1983. Accession 1663.

Trufant, Russell. *The Cranberry Almanac,* 1940.

Slow to the New

Annual Report. New England Cranberry Sales Company, 1909-1949.

"Carver Cranberry Picker Gets Workout," *The Standard Times.* New Bedford, MA, 28 October, 1946.

Cole, Lawrence, and Ruth Cole. North Carver, Massachusetts. Interview, Spring 1983. Accession 1660.

Cranberries. Edited by Clarence J. Hall. Wareham, MA: Wareham Courier, 1936–1966.

Cranberry World, December 1946.

Crowell, William E. Dennis, Massachusetts. Interview, Spring 1983. Accession 1659.

DeGray, R. Cranberry Gatherer. Patent No. 146,579. U.S. Patent Office, Washington, D.C.

Demoranville, I. E. "Frost Forecasting and Frost Protection for Cranberries." Cranberry Station, E. Wareham, MA.

Gomes, Doris. Marion, Massachusetts. Interview, Winter 1983. Accession 1648.

Hall, Clarence J. "Are We Slow to the New?" *Cranberries.* March 1951.

"Harness Helicopters to Cranberry Bogs," *Boston Post,* May 1944.

Jenney, J. R. Cranberry Picker. Patent No. 408,414. U.S. Patent Office, Washington, D.C.

Kiernan, Richard. Wareham, Massachusetts. Interview, Winter 1983. Accession 1646.

Nash, Jean, and Neil E. Stevens, "The Development of Cranberry Growing in Wisconsin." *Wisconsin Magazine of History,* March 1944.

Peterson, Byron S., Chester E. Cross, and Nathaniel Tilden. *The Cranberry Industry in Massachusetts.* Bulletin 201. Massachusetts Department of Agriculture, 1968.

Trufant, Russell. "Economics of Cranberry Picking Machine." *Cranberries,* February 1942.

———. "Equipment Buying." *Cranberries,* April 1955.

——— "The Advantages of Laziness." *Cranberries,* February 1949.

Modern Times

Carson, Rachel. *Silent Spring.* Boston: Houghton Mifflin Co., 1962.

Cranberries. Edited by Clarence J. Hall. Wareham, MA: Wareham Courier, 1936-1966.

"Cranberry Growers' Thanksgiving Recipe: Gather Crop, Oil Well," *Wall Street Journal,* 21 November 1963.

Cross, Chester E. "Recent Activity in the Massachusetts Cranberry Business," *Economic Botany,* Vol. 17, No. 14, 1963.

———. "The Case for Water Harvesting," *Cranberries,* August 1966.

Geisler, Melissa. "Cranberries Profile," Agricultural Marketing Resource Center, February 2009.

Graham, Frank, Jr. *Since Silent Spring*. Boston, MA: Houghton Mifflin Co., 1970.

Henahan, John F. "Whatever Happened to the Cranberry Crisis?" *Atlantic,* March 1977.

"It's Cranberry Season, On Table and in Court," *New York Times,* 28 November 2002.

McPhee, John. *The Pine Barrens.* New York: Farrar, Straus & Giroux, 1968.

Modern Art of Cranberry Cultivation, The. Publication No. 112. Cooperative Extension Service, University of Massachusetts.

"Ocean Spray: The Recent Years, 1960–1980," *Harvest,* Volume II, No. 4, Spring 1981.

Peterson, Byron S., Chester E. Cross, and Nathaniel Tilden. *The Cranberry Industry in Massachusetts.* Bulletin 201. Massachusetts Department of Agriculture, 1968.

Report of Survey: Cranberry Canners, Inc. Booz, Allen & Hamilton, 1945.

"The Surplus Is Gone," *Wicked Local Wareham,* Wareham, MA, 1 November 2007

"Turning Bogs into Villages," *Boston Globe,* 1 April 2007.

Whorton, James. *Pesticides and Public Health in Pre-DDT America.* Princeton, NJ: Princeton University Press, 1974.

Dry harvest, Swan Holt, Carver, Massachusetts.
Lindy Gifford photograph

Index

A

B

C

D

E

F

G

H

I

J

K

L

O

P

Q

R

S

T

U

V

W

Y